The AIRSHIP ROMA
DISASTER *in*
HAMPTON ROADS

The AIRSHIP ROMA DISASTER *in* HAMPTON ROADS

NANCY E. SHEPPARD

THE
History
PRESS

Published by The History Press
Charleston, SC
www.historypress.net

First published 2016

Manufactured in the United States

ISBN 978.1.46711.920.7

Library of Congress Control Number: 2015953420

This book is dedicated in the loving memory of the forty-five men on board ROMA *on February 21, 1922.*

CONTENTS

CONTENTS

INTRODUCTION

In 2012, I found myself looking through the digital collections available online through Norfolk Public Library's J. Sargeant Memorial Collection. Being perhaps the most comprehensive collection of imagery available online for the history of Hampton Roads (Virginia), it is a fascinating window into bygone days of my hometown.

I came across a black-and-white photograph that intrigued me. It was a picture of an airship flying over Granby Street in Norfolk. I had never heard of dirigibles in Hampton Roads before. I asked family and friends if they had ever heard of such a story. Much to my astonishment, they hadn't, either. I was captivated.

The more I researched the dirigible, the more I was drawn to this story. The airship was named *ROMA*, and she met great disaster in Norfolk in 1922. She was a goddess of the skies; the premier dirigible for the U.S. Army Air Service. Within her keel, she housed the best and brightest balloon men the service had to offer.

I was drawn to not only *ROMA* but also to her crew. Each of the forty-five men had vibrant stories, and the hole that they left in the wake of the disaster was vast. However, their stories were quickly buried and forgotten and their memories never truly honored. Perhaps it was my own ties to the military community, but I couldn't help how heartbroken and enraged I felt. For the last four years, I have devoted my time and attention to telling not only *ROMA*'s story but also the stories of her crew.

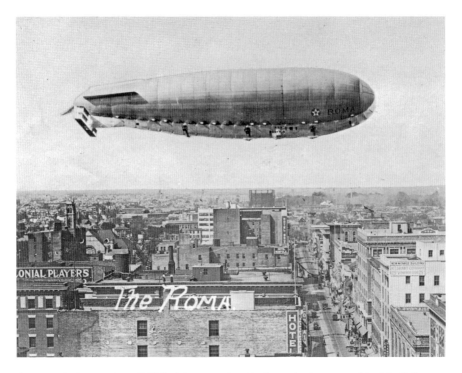

A composite image shows *ROMA* flying over Granby Street in downtown Norfolk. *J. Sargeant Memorial Collection, Norfolk Public Library.*

What you're about to read is not a story simply about an airship or a disaster. This is *their* story—the story of forty-five brave men who took to the skies on February 21, 1922, and were forgotten in the weeks, months and years to follow.

The events you're about to read have been collected through extensive research from personal letters, essays, articles, official testimony, interviews, imagery, genealogy and notations from archives and libraries. Most of the dialogue and descriptions were taken from these sources. However, educated assumptions and a little creative license were taken in order to complete the narrative. It is important to note that very few of the events and conversations have been surmised, and those that were are based on factual evidence. What you're about to read is a forgotten tale that has never been told this comprehensively before.

It is my hope that these words will honor these brave airmen and give them the peace and everlasting memory that they deserve but have never been granted.

CHAPTER 1

THE PERFECT DAY TO FLY

These Italians never do things by halves…they can be very, very good,
especially at entertaining.
—Kenneth L. Roberts[1]

March 15, 1921. The sun shone bright upon the ground near
Rome, Italy. A massive, silver aircraft hangar loomed among the
rolling hills, obstructing the the sun's rays. The hangar was around five
hundred feet long and at least one hundred feet tall. It sat in an empty
field with only a light stone pathway leading to the entrance. Its two
impressively large doors were shut, and a bustle of noise could be heard
from inside the thin corrugated walls. A small crowd eagerly waited on
the grounds surrounding it, their voices chirping with excitement as they
prepared for that day's festivities. Men were dressed in pressed suits and
starched military uniforms, and their female companions wore their best
Edwardian finery. The Italian government declared that this day would
be the perfect one to fly.

Three U.S. Army Air Service officers stood together, their drab green
wool Rickenbacker jackets freshly pressed and their russet-brown knee-
high boots free from scuffs. Major John Gray Thornell, the commanding
officer of his small ensemble, examined every detail of the day's events
with his stoic, deep set eyes, his brow furrowed in pensive thought. He
twisted his small mouth into his infamous cocky half smile, topped by

Wedding photo of Lieutenant Walter J. Reed and Maria Blackiston. Captain Dale Mabry, immediately behind Maria, served as best man. *Hampton History Museum 1987-18-124.*

his aviator's moustache, which sat nestled beneath his prominent yet handsome nose. Underneath his uniform cap, his brown hair was slicked in the fashion of the day. He was the penultimate embodiment of a lighter-than-air aviator and appeared fit to command his crew. His wife, Marie, stood near him, her sweet smiling face framed by the collar of her heavy dark coat and her burgundy cloche hat. Marie was happily chatting with another officer's wife, Mrs. Maria Blackiston Reed.

Maria was the wife of Lieutenant Walter J. Reed. She was a handsome woman with beautiful, narrow features; piercing blue eyes; and a petite frame. Her long, slender fingers pointed to various landmarks and people, and her boisterous laugh and playful personality were infectious. Her husband stood nearby, chatting with his fellow officers. His face was young and friendly, and excitement streamed through his eyes. Lieutenant Reed was soft spoken, but his sweet manner was just as infectious as his wife's personality. He walked with an air of confidence and nearly had a bounce in his step from the zeal that filled him. Walter and Maria had married the day before leaving for Italy less than a month earlier, and they felt this trip would be a perfect opportunity for a honeymoon.

Captain Dale Mabry stood beside his dear friend Lieutenant Reed. Having served together during the Great War, they had an innate kinship, despite their differing personalities. A steady bachelor, Mabry was confident and had a very distinct wit and an air of of invincibility. He had unique, handsome features with a square jaw, sandy hair and a strong build. This made him popular with young women, and he was known for escorting and teasing different young ladies back at Langley Field, Virginia. He had no intentions in following in Walter Reed's footsteps and settling down any time soon. A former real estate agent who hailed from Tampa, Florida, he was a passionate aviator, particularly about lighter-than-air technology, and never shied away from evangelizing his belief that these great ships were the way of the future. [2] Mabry and Reed developed their bond while attending balloon school together in France during the war, becoming so close that Mabry stood as the best man at the Reeds' wedding in New York City on February 3, 1921. The next day, the three officers, along with their two wives and five enlisted men, were en route to Italy to inspect and accept the army's latest purchase: a large, grand airship.

From left: Lieutenant Reed, Major Tornell, Major Chaney, Maria Reed, Marie Thornell and Captain Mabry pose for a photograph in front of *ROMA*. *Fabio Iaconianni.*

On board the transport ship *Somme*,[3] the trip over to Italy hadn't been an easy one. The American crew and their companions shared the tight corridors with a group of Graves Registration officers who had loaded the vessel's hold full of wooden coffins. When a fierce winter storm arose in the Atlantic, Lieutenant Reed advised Maria that if anything should happen to the ship, she was to climb into one of the coffins to float to safety.[4] This made the young bride very thankful when the ship reached its destination of Antwerp, Belgium, on February 28, 1921.[5] That night, the crew and their companions were given special permission to tour the streets of the city. They soon left by train and arrived in Rome by March 4.[6] It became a waiting game before the Italians felt the weather was appropriate for an inspection flight.

The Italian government placed very little emphasis on the actual military inspection of the airship, promising much to their guests. It would be an elaborate spectacle of elbow-rubbing with the elite, wine drinking and viewing the stunning vistas and monuments of southern Italy from the air. On the day of the event, reporters gathered, and photographers quickly wove through the crowd to capture the smiling faces of those about to board. A souvenir photo album was promised to all the guests that day. The eager Americans couldn't help but feel enthralled by the perfect weather, unspoiled landscape and exclusive company that they were going to keep that day.

An older gentleman with a cane approached the group. His glasses seemed to prop up his black homburg hat, and a full, bushy white beard framed his face. He had a pleasant smile, soft eyes and a gentle step as he approached. He introduced himself as Robert Underwood Johnson, the American ambassador to Italy. He was a bit of a celebrity, not only for his political position but also because he was an acclaimed writer and gifted poet. Flanked by men working for his office and his granddaughter, Olivia,[7] Ambassador Johnson had a stately yet humble air about him.

Airship engineer and designer Umberto Nobile chatted politely with Ambassador Johnson and the American officers. He was a thin man with high cheekbones and a small moustache that accentuated his sharp features. He wore thin silver glasses and a dark suit.[8] He was never reluctant to discuss the innovative lighter-than-air design and engineering at Stabilimento di Costruzion Aeronautiche in Rome, Italy—the central location of aircraft design and construction for the Italian government.[9]

Gathered on the opposite corner of the field were the five enlisted men whom the officers had personally chosen to accompany them on

Right: (Left to right, top row) Sergeants Biedenbach, Hoffman and Beall and Master Sergeants Chapman and McNally along with (bottom, left to right) Lieutenant Reed, Major Thornell and Captain Mabry in Italy. *Library of Congress.*

Below: American ambassador to Italy Robert Underwood Johnson (left) and *ROMA* designer Umberto Nobile (right) before *ROMA*'s inspection flight in Italy. *Fabio Iaconianni.*

ROMA's noncommissioned officer crew in Italy. (Back from left) Sergeant Biedenbach, Sergeant Hoffman, Sergeant Beall and Master Sergeant McNally. Master Sergeant Harry Chapman sits in front. *Air Combat Command History Office, Joint Base Langley-Eustis.*

this voyage. They were the five best engineers and crewmen that the air service had to offer. The men stood in a circle, attempting to make small talk with a few members of the Italian crew, though a language barrier inhibited any true interaction. Nonetheless, it was a pleasant moment and allowed all of the men to unwind before a long day ahead. Unlike their superior officers, the crewmen's day was not to be filled with socializing but rather giving the airship a thorough inspection from stem to stern.

The leading crewman was Master Sergeant Roger McNally. At thirty-four years old, he was the oldest and the only married man of the enlisted crew. When the Great War erupted, he was teaching school for the Department of the Interior in a remote part of Alaska. Upon learning the news, he walked four hundred miles to the nearest recruiting station in Seward.[10] He was an electrical engineer by trade and considered to be among the most competent in the U.S. Army Air Service.[11] He tried not to feel self-conscious of the age difference between him and the other men, though it was perfectly obvious that he occasionally felt uncomfortable around boys a decade or more his junior. Roger slicked his dirty blond hair back to cover a small bald spot appearing on the back of his head. He stood with impeccable posture, and his uniform was always perfectly crisp and situated. Roger McNally had a quiet and reserved personality but nestled beneath was a wealth of knowledge and life experience.

The next senior enlisted man in the company was Master Sergeant Harry Chapman. He joined the army in 1916 and had attended balloon school at Fort Omaha, Nebraska, after which he served during the war in France as part of the Twenty-fifth Balloon Company.[12] Loyal to a fault, he upheld the standards and expectations of a soldier with his every being. His tight facial features often looked pinched when he was concentrating, and he had a slight under bite, making his chin appear particularly prominent. His serious demeanor and downturned eyes made him seem perpetually sad.

If Chapman was the serious member of the group, then Staff Sergeant Marion "Jethro" Beall was his foil. Jethro had a witty personality and rarely shied away from saying what was on his mind. He was one of the first men in Macon County, Missouri, to enlist for World War I, though he would not leave stateside to serve. Instead, Jethro was assigned as a motorcycle messenger in the Motor Transport Corps and then later as a White House guard.[13] He was discharged in 1919 and immediately reenlisted in the U.S. Army Air Service.[14] He was a tall and lanky man with a thick mop of dark brown hair that he attempted to slick back, a particularly strong jawline and a distinct smile.

The other affable member of the enlisted ensemble was Sergeant Virgil Hoffman. A native of Eaton Rapids, Michigan, Virgil enlisted in the army in 1918, serving with the Third Balloon Company in France. At only twenty-two, he had a winning laugh and bright smile. Though his eyes conveyed

Sergeant Virgil C. Hoffman poses for a portrait. *Hampton History Museum 1987-18-132.*

seriousness, he walked with a bounce in his step and carried his kind heart on his sleeve. He was shorter than the others, with a wave of light brown hair. His uniform often appeared slightly rumpled though still well appointed. Virgil's eager attitude and happy personality won over everyone he came into contact with, making him one of the more well known and liked of the enlisted men assigned to Langley Field, Virginia.

Sergeant Joseph Biedenbach poses for a photograph. *Hampton History Museum 1987-18-119.*

Each of the four men shared the experience of serving during the Great War. However, the youngest member of their party, Sergeant Joseph Biedenbach, did not. With the postwar draw down of the armed forces, there was a push to enlist more men into the lighter-than-air division of the U.S. Army Air Service. Joe readily enlisted and proved himself competent enough to be promoted to sergeant within just one year.[15] He had a very youthful face that made him look more newspaper boy than able and trusted balloon man—a position he absolutely loved.

Cameras filming newsreels rolled, and photographers snapped photographs with their accordion lensed cameras. Then, 160 Italian enlisted men peeled open the large doors of the hangar, which screeched with the sound of ungreased metal rubbing against itself, and the echo of this sound resonated from within. With a loud bang from the doors hitting the outside walls of the corrugated building, an enormous, lattice metal hemispherical frame was revealed. What followed behind was lost as the sight was swallowed by the cavernous hangar. The onlookers curiously glanced inside as best they could to no avail. The handling crew retreated into the shadows, disappearing for a few minutes.[16]

They reappeared, pulling vigorously on long, thick ropes tethered to their hands. Then, slowly emerging from the cocoon of the hangar, was a silver, glistening airship. The men walked the beautiful machine into the sunshine, and a collective gasp of wonder came from members of the crowd as they gazed upon the 410-foot-long airship. Painted in black seven-foot-tall letters on the side of the envelope was her name: *ROMA*.

CHAPTER 2

THE GREAT AIRSHIP *ROMA*

*I am most impressed with the semi-rigid type of airship and
believe it has great possibilities.
—Major John G. Thornell*[17]

A long with the seven other men in his company, Major John Thornell was charged with bringing *ROMA* to the United States. She was to become the first large airship in the air service's balloon fleet. German zeppelins proved a formidable foe during the the Great War. They were less fragile than their fixed-wing, or "heavier-than-air," counterparts. Throughout the war, airplanes proved themselves fragile machines, with the average lifespan of their pilots being as little as eleven days. By the end of the war, the U.S. Army Air Service estimated that only 2,200 of its 10,000 planes were still in serviceable condition.[18] Zeppelins were an initial advantage to the German military, which used them for strategic bombing purposes.[19] With Germany defeated, the Treaty of Versailles demanded for the Rhineland to be demilitarized, including its air force. The airships were sent to Great Britain with some given to the United States.[20] Lighter-than-air craft could fly higher, carry more cargo or people, stay aloft longer and travel much greater distances. Airships were often regarded as the future of aviation. In a world that looked to the skies, there was even chatter of airships taking the place of battleships one day. The U.S. Navy had been granted the responsibility of the use of massive zeppelins, while the army was resigned to small free balloons and blimps that were merely used for observational purposes. When the navy

(Back from left) Sergeant Virgil Hoffman, Sergeant Joseph Biedenbach, Sergeant Beall, Master Sergeant McNally, Master Sergeant Chapman, (front) Lieutenant Reed, Major Thornell and Captain Mabry. *Air Combat Command History Office, Joint Base Langley-Eustis.*

passed on the opportunity to purchase a used but innovative airship from Italy, assistant chief of the U.S. Army Air Service, Brigadier General William "Billy" Mitchell, lobbied the War Department to purchase that same ship for the air service. This new acquisition was *ROMA*.[21]

American ambassador to Italy Robert Underwood Johnson and American and Italian officers prepare for *ROMA*'s inspection flight in Italy. *Fabio Iaconianni.*

The crowd gathered in a line to board the ship, lunch bags in hand. They watched as two strong men from the Italian army hurried to remove bags of sand ballast from the front of the ship to counter the extra weight of the passengers. Members of the Italian military, Umberto Nobile as well as *ROMA*'s other designers, Celestino Usuelli, Colonel Arthur Crocco and Pissone flocked around the Americans to tout *ROMA*'s attributes.

One by one, passengers boarded through a small hatch in the center of the ship. They walked through a control room onto a boardwalk and into a well-appointed passenger cabin. With limited seating, passengers politely conversed over who would sit where as their lunch bags were loaded into the space that once held the ballast. Prince Viggo of Denmark came into the cabin, sporting a light blue military uniform covered in gaudy accoutrements that dazzled the other passengers. Each clamored for an opportunity to be seen with royalty. The American military guests retreated to the stations of their own personal familiarity for observation.

The Italian engineers walked out on six precarious metal scaffolding that acted as catwalks to each of the Ansaldo engines, and a deafening buzz filled the air as they were started. The handling crew members loosened their

Lieutenant Walter J. Reed glances back from one of *ROMA*'s observation windows during *ROMA*'s inspection flight in Italy. *Fabio Iaconianni.*

grip on the ropes that tenuously held the mighty ship to the ground. Out of nowhere, Captain Mabry appeared in the doorway of the passenger cabin and rather exuberantly exclaimed, "Hey, we're going up!"[22]

Guests shuffled and pushed for a coveted spot at one of the recessed windows in the passenger cabin as the distance between themselves and the ground grew. They began flittering between the passenger and control cabins to get a better view from large observational windows in the control cabin. Major Thornell, Captain Mabry and Lieutenant Reed watched with unparalleled attention and zeal as the Italian officers maneuvered the controls. They often asked questions and took turns at *ROMA*'s various control stations. While, in theory, very similar to the small pony balloons[23] the birdmen (an affectionate nickname for aviators) were used to flying, this ship was a brand-new form of aviation for them. This prospect fascinated and excited the three officers.

It seemed a very short time before the ship leveled off and the passengers' excitement seemed to mellow. Ambassador Johnson pulled a notepad and pencil from the recesses of his coat and began taking notes of his observations, in no doubt for poem he would compose later on. Marie Thornell sat with Olivia Underwood, and the two retrieved knitting needles and yarn from their purses. They went to work quietly, their fingers moving with incredible ease and speed. The volume of conversation among the passengers exceeded polite conventions in a fruitless endeavor to overcome the deafening sound of the engines.

One of the Italian passengers glanced around at the rather unimpressed expressions on the faces of the other passengers. He took it upon himself to become the self-appointed tour guide for the trip. In a booming voice, he announced, "In America, you have ceased to drink. So it is possible that these places below us will have no interest for you. Over there against the hill slope, with the many large white buildings, is Frascati. The wine of Frascati is sold largely in Rome. It is a pleasant wine, neither too sweet nor too sour, but rather heavy. Beneath the town there are great caves where the wine is stored, and it is very pleasant to go into these cool caves on a hot day in the summer and sample the contents of the casks. The sampling is done with small glasses fastened to the end of long sticks, and the guardians of the caves thrust the glass first into one cask and then into another cask, insisting that the visitor drink a glassful from each other. After a few glasses one suddenly knows what it is like to be kicked by a mule."

Virgil Hoffman happened to be passing through the cabin to catch the tail end of the lecture. He licked his lips and commented, "I have a bottle of Frascati in my lunch bag."[24]

Jethro Beall made his way behind Virgil. His face had an expression of pleasant amusement as a wide smile spread across his face. The Italian gentleman rolled his eyes and continued on his lecture. "As for the wine of Orvieto, the bottles in which it is sold always bear the words, '*Est! Est! Est!*' to commemorate the excellent taste of the German bishop. But, of course, such things do not interest you Americans, who live in a country that frowns on drinking."

Jethro scoffed under his breath and retorted in a hoarse voice, "I have two bottles of that Orvieto stuff in my lunch bag. We might as well get 'em out and frown on 'em!"[25] The cabin burst into a roar of laughter, and the tour guide mumbled to himself as he retreated to a corner of the cabin.

A journalist for the *Saturday Evening Post* named Kenneth L. Roberts walked about the cabin, making a point to speak to every single person. His jovial smile beneath his brown fedora and his pleasant demeanor made it easy to open up to him as he documented the trip for posterity. A photographer snapped pictures while the hungry guests dug their hands into their recently retrieved lunch bags. Passengers took turns exchanging seats and milling about the cabins. The American crew drank along with their Italian hosts, explaining that it would be impolite to decline the invitation to do so, but secretly relishing the opportunity. Major Buffi, the Italian commanding officer, allowed passengers to take turns steering the ship, with the wheel often left unattended by the Italian crew. The Italians were brilliantly executing their plan to ensure the atmosphere was more of a social engagement than actual military operation. Kenneth Roberts scrawled on his notepad, "These Italians never do things by halves…they can be very, very good, especially at entertaining."[26]

The stillness of the flight was abruptly interrupted by a jarring roll of the ship. Maria Reed sat down on the floor of the cabin, her face turning unnaturally pale from nausea. She desperately gripped her stomach while attempting to regain her composure. Along the side of the ship, the glassless portals were quickly occupied by passengers with weaker constitutions. One Italian lady's face became pallid as she rushed to the windows in a fruitless attempt to find a space to lean her head from. Finding there were none, she hurried toward the control cabin to lean out of one of the large observational windows. Finding the catch of the door jammed, she frantically began rattling the handle. Crew members gallantly rushed to her side, also failing at their own attempts to release the door's latch. Virgil searched through the various crevasses in the cabin and retrieved a screwdriver. He swiftly dove in and ferociously attacked the latch. Seeing the increasingly sick expression on the woman's face, the other passengers retreated to the opposite end of the cabin, fearing the scene that might unfold if she were not able to find a window in time. After Virgil removed the three screws of the latch, it fell from the door and an Italian officer rushed the woman into the control cabin.[27]

Slowly, the rumblings ceased and seats were happily relinquished to those passengers who had experienced the worst of the air sickness. The mood once again settled into one of quiet enjoyment.

Jethro Beall made his way through the ship toward one of the engine doorways. With only metal framing and taut fabric separating him from the void below, it felt almost like he was walking through a camping tent rather than the gondola and keel of a large airship. One of the Italian engineers waved for him to join him. Jethro smiled and waved back. The engineer wrapped a harness around Jethro's waist and attached a tether to him. "So you won't fall," the engineer said, his broken English and heavy Italian accent difficult for the man from Missouri to understand. Jethro smiled and stepped toward the doorway. A slender board sat nestled in metal scaffolding with only thin handrails to hold on to. The roar of the Ansaldo engines filled the air, and the wind whizzed past Jethro, making it difficult for him to keep his balance. He glanced at the massive eight-foot propeller rapidly spinning on the back of the engine. He knew that if he lost his balance, he would either fall into the arc of the propeller or plummet to the ground. Jethro didn't quite trust the cable that tethered him to the ship. However, it was considered dangerous to wear to a parachute this close to the engine in open air.[28] There wasn't room for fault or fear. Though he was there to learn about the Ansaldos, he couldn't help but steal a moment to look down on southern Italy in the way only a bird would. It was a beautiful sight that momentarily took his breath away. He briefly forgot that he was standing more than five hundred feet above the ground with only a harness and cable keeping him from falling to the Earth.

Looming in the distance, Jethro spotted the foreboding sight of Mount Vesuvius. Inside the ship, the passengers rushed to the windows to catch a glimpse of the infamous volcano. It had been their understanding that the itinerary would include passing over the summit to look down into the caldera. There was momentary confusion as *ROMA* turned port to avoid flying directly over the volcano. Never missing a beat, Kenneth Roberts commented on his perplexity over the sudden change. Another guest, who claimed to be knowledgeable on lighter-than-air technology and to have an intimate affiliation with *ROMA*, replied, "The life of a dirigible's gas bags are only about two years. *ROMA*'s have been inflated for at least fifteen months…they're leaky and probably couldn't get enough lift to get over Vesuvius. After all, this flight is merely the Americans' one and only demonstration flight. Since she has been wallowing so badly on such a cloudless and windless day, Major Buffi probably decided to pull her over the cool sea instead of the hot ground so she would not stand on her nose and then her tail."[29]

Commentary and suspicions over trouble were conveniently interrupted by a man entering the passenger cabin. He was dressed as a waiter and moved swiftly to clear space. He disappeared for a moment only to reemerge with a folding table. He spread it and laid out a rumpled linen tablecloth. Then he retrieved a large hamper filled with fresh china, silverware and an assortment of food and wines. The waiter turned the chairs in the cabin toward the table and placed menus at each place setting. Ambassador Johnson was the first to approach the table. He picked up of the menus and read it aloud: "Cold meats, fish in jelly, fillet of veal, pastries, coffee, Chianti, Marsala and white wines!"

Right on cue, Major Thornell, Captain Mabry and Lieutenant Reed appeared in the cabin, taking their seats at the table. A toast was given to the friendship between the two countries. Ambassador Johnson followed with one of his own, speaking to the genius of the designers. Signor Usuelli thanked Ambassador Johnson and announced that he would soon begin designing another semi-rigid ship, saying, "This one will be four times the size of *ROMA*!"[30]

Alcohol flowed freely throughout the meal, making gregarious laughter and tipsy mannerisms commonplace among the guests. Occasionally, crewmen were enlisted to walk out onto one of the engine platforms and toss an empty wine bottle to the ground below. This was a rather perilous place to be considering the intoxicated state of those on board. Major Thornell leaned back in his chair and commented to his wife, "I sure do miss my after dinner cigar!"

Colonel Crocco replied, "No, no! A match lit near the hydrogen is very dangerous. Any spark near it could detonate a large explosion!"

Mabry scoffed, waving one of his hands in the air. He remarked, "One could light a bonfire on the floor of the cockpit and cook a meal over it without the slightest danger of an explosion!"[31]

As *ROMA* drew closer to her hangar on the hills near Rome, souvenir bottles of champagne were passed out to the guests, many of whom chose to immediately indulge themselves. Since there weren't facilities to house dirty dishes, guests piled them around the cabins and playfully tossed scraps and trash out the windows. There was more laughter and merriment than serious observation, and the Americans easily fell for the hospitality of their Italian hosts. However, they hadn't been able to garner a complete understanding of the technology they were purchasing on behalf of the air service.

After eight hours in the air, *ROMA* was pulled to the ground and the passengers stumbled from the ship. Captain Mabry approached Kenneth Roberts and patted him on the back. In a bold, almost drunken tone, Mabry proclaimed to Mr. Roberts, "Come and take a ride with us when we get this boat to America! We'll run her like a battleship…sail in any weather! We'll work the men in shifts; four hours on and four hours off. None of this all-day steering stuff and only flying in 'picnic' weather!"[32]

Despite nursing a wine-induced headache, Major Thornell managed to send a quick telegram back to the air service's top brass. It read:

> *Aboard the airship* ROMA *during an inspection trip: I consider* ROMA *a truly wonderful ship, easy to handle, able, very comfortable and complete. The inspection trip surpassed expectations. The ship has good speed and ought to be a great success as a commercial airship over land or water.*
>
> *I am most impressed with the semi-rigid type of airship and believe it has great possibilities. The Italian engineers who designed and built the* ROMA *deserve great distinction for accomplishing such an engineering feat.*[33]

Chapter 3

ROMA AND HAMPTON ROADS

*Though the semi-rigids are not built…in the United States, the training of
American airmen in handling this type will be valuable to the service…
—Major John G. Thornell*[34]

There was no doubting that *ROMA* was a marvel of a ship and unlike any of her contemporaries. She was 410 feet long and 92 feet tall, dwarfing all other ships in the air service's fleet. *ROMA* had a gas volume of approximately 3,400 cubic meters[35] (1,193,000 cubic feet) of hydrogen lifting gas. She had the model number "T-34," standing for both the ship's volume and "transatlantic."[36] Originally meant for military purposes, the Italians were forced to repurpose her into a commercial travel vessel at the end of the Great War. After taking her first glorious flight over the Italian hills on March 19, 1920,[37] the Italian government made grand plans to make transatlantic flights between Rome and Rio de Janeiro.[38] However, these flights would never come to fruition, and she was just used for sightseeing purposes over southern Italy. For reasons they would not disclose, the Italians deduced that it would be a better economic decision to sell *ROMA* to the Americans rather than keep her in their service. To the U.S. Army Air Service, *ROMA* was a treasure. For the Italians, ridding themselves of the airship was like a silent sigh of relief.

Classified as a semi-rigid airship, *ROMA* was an engineering masterpiece. The larger zeppelins kept their shape by way of a skeletal structure inside the airbag, thereby granting them the classification "rigid airships." Their much

The small army airship *A-4* sits next to the partially reassembled *ROMA* in her hangar. *Air Combat Command History Office, Joint Base Langley-Eustis.*

smaller non-rigid blimp cousins lacked rigid structure and could only carry a limited number of passengers. *ROMA* was an interesting hybrid of the two designs. While her long, cigar-shaped envelope lacked rigid framework, her triangular keel within the confines of the bag had an articulated rigid structure. Located inside the keel were the control and passenger cabins. Attached to the front of the rigid keel was a large, latticework copula (or nose cone) that fit perfectly on the front of the envelope.[39] The copula maintained the structure of the forward section and measured approximately forty-five feet in diameter.[40]

Inside the envelope, eleven ballonets[41] containing hydrogen maintained the shape and structure of the envelope. Each ballonet's pressure and purity was closely monitored in the control cabin through a series of eleven glass manometer tubes, which showed the pressure within each of the cells. Because of the semi-rigid design of *ROMA*, it was necessary for the internal pressure within the envelope to be maintained for the ship's structural stability.[42]

Maintaining purity was of vital importance due to the rather unstable nature of hydrogen gas. The purer the hydrogen was within the ballonets, the more difficult it was to ignite. However, when oxygen mixed with hydrogen, thus dropping the purity of the hydrogen, the gas would become unstable and would be more likely to combust.[43]

ROMA was capable of carrying a substantial amount of ballast that consisted of both water and sand distributed in various locations throughout the ship.[44] She also had several gasoline fuel tanks, each one measuring approximately two feet in diameter and nearly four feet in height.[45] The tailfins were a typical cruciform shape, but attached to the back of the keel was a large triplane rudder, often referred to as a "box-kite." This was an incredibly unique structure, allowing for greater ease of movement and elevation in the skies.[46] *ROMA* was absolutely exceptional, innovative and exciting for her new American crew.

The six engines on board *ROMA*, Ansaldo twelve-cylinder, type 4 E model 2940,[47] were unfamiliar to the American engineers. These Italian-designed

Members of *ROMA*'s crew stand with the reassembled box kite rudder assembly prior to its installation. *Air Combat Command History Office, Joint Base Langley-Eustis.*

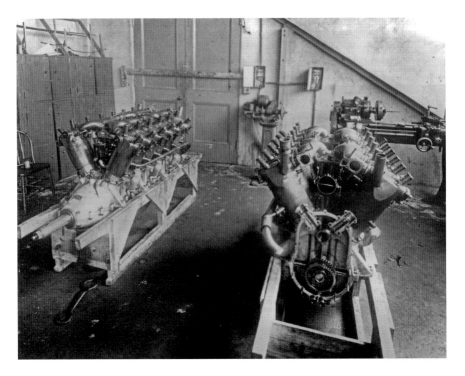

ROMA's Ansaldo motors during their reassembly after being shipped from Italy. *Air Combat Command History Office, Joint Base Langley-Eustis.*

engines were heavy and capable of 450 horsepower.[48] Each pair of engines were aligned so the forward pair angled at an incline of twelve degrees, the center at ten degrees, and the aft were parallel to the ship's keel in order to avoid a slipstream effect on the ship while in flight.[49] Each engine had two propellers: a tractor and a pusher, allowing for greater ease of movement and landing. However, the propellers required midair changes—quite a dangerous feat on the scaffold engine platforms that extended freely on the sides of the ship.

None of the American engineers had ever worked with Ansaldos before and questioned their capability of withstanding the temperature variations quite common in the American Mid-Atlantic climate.

With *ROMA*'s one and only inspection flight behind them and Major Thornell's rave reviews arriving to the brass back home, the American and Italian governments finalized their contract to purchase the behemoth. The Italians originally asked for a purchase price of $475,000 but quite readily negotiated for the United States government to pay the paltry sum of $184,000.[50] During negotiations, Thornell repeatedly attempted to add provisions to the contract for purchasing a new envelope for *ROMA*. Secretary of War John W. Weeks denied this addition, saying that one would only be appropriated when it was deemed absolutely necessary.[51] When Major Thornell relayed the news to his crew, Captain Mabry scoffed, "This envelope will only last three or four months longer!"[52] It wasn't unusual for airships to fly with an envelope that had some age as long as the bag was able to maintain hydrogen purity within the ballonets.

In a show of affection between the two countries, the Italian Crown bestowed on Major Thornell, Captain Mabry and Lieutenant Reed the Order of Knighthood of the Crown of Italy in late April.[53] Newspapers all over the world were abuzz with this grand gesture.

Major Thornell then advocated flying *ROMA* back to the United States, using her for the transatlantic capabilities the Italians once intended.[54] It was theorized she was capable of traveling the distance while carrying a substantial amount of weight or passengers. However, Thornell's request was quickly denied, and the eight American men began overseeing the process of disassembling the great ship and packing her into wooden shipping crates.

Lieutenant Reed was put in charge of the Americans and Italians packing *ROMA*.[55] Each piece of the articulated keel, rudder, copula and envelope were carefully inspected before being packed away for shipment. Each crate was meticulously inventoried and labeled. Lieutenant Reed diligently inspected each section as the ship was taken apart, ensuring every piece was tagged and marked so that it could be properly reassembled once the crates reached America.[56]

Day and night, the men worked disassembling their new ship. Lieutenant Reed studied the Italian blueprints with a keen eye while Jethro Beall and Virgil Hoffman busied themselves with studying the mechanics of their new engines. Roger McNally was to be *ROMA*'s shipmaster.[57] He inspected every inch of the articulated keel, the ballonets and the silk envelope. Harry Chapman and Joe Biedenbach helped wherever they could, learning every facet of the ship. Though free time was scarce, the crew and their companions were able to take in a great many sights. Their Italian hosts were generous in sharing not only their knowledge and expertise with the Americans but also their company and merriment.

After eight painstaking weeks,[58] the large wooden crates containing *ROMA* were transported to Genoa and then loaded onto the U.S. Navy collier *MARS*.[59] *ROMA* wouldn't again see the light of day until she arrived at Langley Field in southeastern Virginia. The Americans bid a fond farewell to their generous Italian hosts and headed back to the United States.

Langley Field sat nestled where the James River, Elizabeth River and Chesapeake Bay meet in southeastern Virginia. The post was surrounded by the slowly emerging communities of Hampton and Newport News, both cities part of a region better known as Hampton Roads. The locals clung tightly to their agrarian roots yet were slowly beginning to embrace a vibrant tourism industry as throngs of vacationers started flocking to the area for its pristine beaches and ideal weather.

This was a region deeply rooted in American military history. The great Civil War ironclad battle of USS *Monitor* and CSS *Virginia* had occurred in the James River,[60] and Fortress Monroe was one of the oldest active forts in the country. It had been the only Union stronghold in Virginia during the Civil War and was a refuge for escaped slaves.[61]

Just across Hampton Flats,[62] the U.S. Navy sailed the "Great White Fleet" from the Elizabeth River in Norfolk during the Jamestown Exposition of 1907. The Jamestown Exposition was a World's Fair of sorts held to honor the tercentennial of the establishment of the colony at Jamestown. The exposition was a financial failure, and the grounds were left in ruins until the War Department purchased them and established thriving U.S. Navy bases for both maritime and aviation purposes.[63]

In 1917, the National Advisory Council for Aeronautics (NACA), the predecessor for NASA, decided that there needed to be an airbase that not only sat on flat, low-lying land but was also located near the water and near a navy base for joint operational purposes. Hampton Roads was chosen as the perfect location to establish such a post, and thus Langley Field was born.[64] The post was home to the U.S. Army Air Service Lighter-Than-Air School, Balloon Companies 10 and 19, a photography school, heavier-than-air craft squadrons and a research laboratory for NACA.[65]

The centerpiece of Langley Field was a massive airship hangar. The hangar was recently built and completed in time for *ROMA*'s arrival in 1921. It stood 116 feet tall, 420 feet long and 125 feet wide.[66] The hangar was just mere footsteps from the water on a sunny field with sparse clusters of trees along the shoreline.

A hydrogen generating and processing[67] plant was located conveniently near the hangar. It was a tall brick building with large plate-glass windows. Massive machines echoed within the walls as a bustle of men filled and moved cylinders containing hydrogen to a stuffy wooden warehouse nearby. The cylinders often overflowed the warehouse and subsequently stacked one on top of the other precariously in parallel pathways. Though highly combustible and unstable, hydrogen was the preferred gas to use in airships.

ROMA flies away from Langley Field during one of her U.S. trial flights. *Air Combat Command History Office, Joint Base Langley-Eustis.*

Because of the extremely light nature of the gas, airships were able to achieve greater lift. Hydrogen was also easy to generate, and most important to the government, it was far cheaper than its safer counterpart, helium.[68]

Langley Field was an astounding place on the cusp of securing the U.S. Army Air Service an eternal place in aviation history. In local theaters, newsreels played on silver screens encouraging the best and brightest young men to enlist in the army's lighter-than-air programs. Dale Mabry was a star in his own right, often featured in these news reels, commanding crews of smaller airships in flight.[69] Langley gave these boys a sense of freedom and adventure that they craved, and the balloons were the vehicles to provide what they sought. For the veterans of "LTA," an acronym used to refer to "lighter-than-air," Langley offered an opportunity to impart their wisdom to the new boys seeking to become "birdmen" within the cotton embrace of the air service's balloons.

For men like Dale Mabry, pleasure was also found in the social experience of the area and teasing the flock of local girls who perpetually followed

Sergeant Virgil Hoffman and Stella Hoover share an impromptu embrace while at the beach. *Hampton History Museum 1987-18-127.*

handsome aviators around town. However, many of the boys met local women and fell in love. Langley Field was a place where dreams were made and heroes were born.

For Sergeant Virgil Hoffman, that's exactly what Langley was. Growing up in the small town of Eaton Rapids, Michigan, he came from a family that worked on building the railroads. After serving overseas during the Great War, he was stationed at Langley to finish out his enlistment. During this time, he met a local girl named Stella Hoover. He was absolutely smitten by her large curious eyes, vibrant smile and gentle personality. When his enlistment came to an end, Virgil moved back to Eaton Rapids to work on the railroads with his brother, Earl. On Sunday afternoons, he would retreat to a quiet space in his parents' home and would read and reread the letters that Stella sent to him. He was completely in love with this girl. Virgil made the decision to reenlist in the army only under a guarantee that he would return to Langley Field to be with his beloved Stella.[70] Every chance he had, Virgil would make the long trek from his barracks at the post to the beach community near Hampton named Phoebus, where Stella lived with her parents. The two were seamless, completing each other's thoughts and sharing the other's hopes and dreams. They had planned to wed in July the following year.[71]

On July 7, 1921, a small navy blimp, *C-3*, left the ground at Naval Air Station Norfolk for a photographic mission with six men on board. While the handling crew was still holding on to the ropes, a ripped panel on the envelope tore, allowing the outside oxygen to seep in and mix with the hydrogen. The small airship immediately ignited. The riggers on the ground swiftly pulled the small blimp from the sky, saving all of the crew's lives. The six men escaped with only minor burns and injuries. The instability of hydrogen and the necessity for proper maintenance of an envelope were overshadowed by the heroic acts of both the handling and air crews, as well as the amazement that everyone walked away relatively unscathed. The incident was considered by the War Department as a minor infraction and quickly tucked away.[72]

CHAPTER 4

ASSEMBLING *ROMA* AND THE CRASH OF *ZR-2*

Hydrogen…will still have to be used for a time, and while its use sometimes leads to danger…it will be sufficiently safe.
—Washington Herald, *August 30, 1921*[73]

In early August 1921,[74] the crates containing *ROMA* arrived at Langley Field. After spending two months on board the transport collier, the anticipation to reassemble her was nearly unbearable for her new crew. Newspapers across the country buzzed with headlines of her arrival, raising the airship and her crew to monolithic proportions. Americans flocked to newsstands with bated breath, gorging on every word written about this majestic machine. Locals living near Langley Field traveled to the post on the weekends just to catch a glimpse of the colossal crates housing the grand ship. How marvelous would it be to see such a large ship travel through the skies above Hampton Roads!

ROMA's crew was picked from the best and brightest that the air service had to offer. Among those reporting for duty was Lieutenant Byron T. Burt Jr. During the war, he was assigned in France as an observer in free balloons, which looked like hot air balloons used for observational purposes during the war. During this time, he was shot down by the enemy three times. Burt had been awarded by the air service for ensuring the safety of his crewmates over his own.

Some of the shipping crates *ROMA* was packed into for transport from Italy to Langley Field, Virginia. *Air Combat Command History Office, Joint Base Langley-Eustis.*

In 1919, he was tasked with flying a balloon during an international race over Lake Michigan. During this race, his balloon crashed, and he was left adrift on the lake for half a day.[75] His fellow airmen envied Lieutenant Burt's seemingly unparalleled luck for having survived so many incidents. Having just transferred to Langley Field from teaching at a lighter-than-air school, Burt had an unrivaled understanding of the engineering and physics involved with airships. Burt was ambitious and had a serious demeanor. He was a trusted and knowledgeable airman whom any commanding officer would feel lucky to have in his control cabin.

Also reporting for duty was twenty-three-year-old Corporal Alberto Flores. He was a shorter, young Puerto Rican man with a cheery demeanor, and Corporal Flores easily befriended his fellow crewmates. Being tasked as a rigger on board the great ship, he became a "go-to" man who was instrumental in *ROMA*'s reconstruction. He had a funny personality and jovial smile. The officers and enlisted men alike lovingly nicknamed him "Little Flores."[76]

Left: Lieutenant Byron T. Burt (left) and Captain Dale Mabry stand together with speaking trumpets in their hands. *Air Combat Command History Office, Joint Base Langley-Eustis.*

Below: (From left) Sergeant Thomas Yarbrough, (back) Private Gus Kingston, Corporal Irbey Hevron and Corporal Albert Flores. *Hampton History Museum 1987-18-134.*

ROMA's crew works to rebuild the ship's Ansaldo engines at Langley Field, Virginia. *Hampton History Museum 1987-18-24.*

Lieutenant Reed stepped in to oversee the unpacking of *ROMA*. One by one, the heavy Ansaldo engines were removed from their crates and sent to NACA's facility for testing. After the testing was completed, Sergeant Joe Biedenbach was put in charge of overseeing each engine's overhaul in order to bring them to the air service's standards.[77]

The keel and copula were unpacked and each piece meticulously inspected for damage. A civilian from the Goodyear Company named Charles Brannigan joined Lieutenant Reed in the duty of overseeing the reassembly of the rigid sections of the ship.[78]

News soon reached Langley from their airship colleagues regarding the navy's newest zeppelin, *ZR-2*. In 1919, the U.S. Navy purchased from the British plans for a new zeppelin, *R38*. After paying the British to construct this ship, the U.S. Navy reclassified her *ZR-2*. She was an impressive zeppelin at 695 feet long and could fly as high as 21,120 feet. She hadn't yet flown under American colors so therefore was not yet considered part of the American air fleet.

Several U.S. Navy airmen were sent to England to train on board the airship before she was to join the fleet. With a forty-nine-man crew consisting of both British and American balloon men, *ZR-2* took the skies on August 24, 1921.[79] When oxygen leaked into the envelope and mixed with hydrogen, *ZR-2* exploded midair. Forty-four of the forty-nine men on board

Staff Sergeant Jethro Beall stands next to one of *ROMA*'s Ansaldo engines with its massive eight-foot propeller. *National Museum of the U.S. Air Force.*

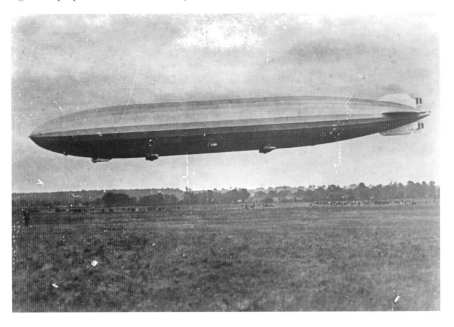

The hydrogen-filled navy airship *ZR-2* takes off from England. She eventually crashed, killing forty-four out of forty-nine men. *Library of Congress.*

were killed. The five survivors were all located in the tail section at the time of the disaster.[80] When news began to circulate of the tragedy, heart-rending cries rose from the public for the War Department to discontinue the use of the dangerous hydrogen gas in American airships.

Lieutenant Clifford Tinker, a publicity officer for the U.S. Navy, accompanied the remains of the American aviators who died on *ZR-2* back to the United States. Having seen firsthand the dangers of hydrogen, he made a point to himself that he would never fly on an airship that was filled with the gas.[81] He also vowed to do whatever he could to convince the War Department to replace hydrogen with a far more stable alternative, like helium. The War Department skirted discussing the dangers of hydrogen and diverted attention from the tragedy to the up-and-coming *ROMA*.[82]

With thoughts of *C-3* and the tragedy of *ZR-2* fresh on everyone's minds, anxiety toward hydrogen began to grow. However, many of the officers attached to *ROMA* continued to insist on the safety of the gas and on the strength of their ship. *ZR-2*'s crash was eventually determined to have been caused by weakening in the structural integrity of her rigid frame.

IT WAS A SWELTERING, HUMID DAY in late August when unpacking began on *ROMA*'s immense envelope and ballonets. The men gathered eagerly in the hangar as Master Sergeant Harry Chapman, Sergeant Virgil Hoffman and Corporal Flores used crowbars to pry open the wooden crate. They reached in and began steadily pulling out the envelope. Sergeant Chapman knelt down, inspecting the fabric in his hands. His face pinched as he called, "Lieutenant Reed, take a look at this, sir!"

Walter Reed made his way over to Chapman. There was a large rip in the envelope over gas compartment 9 and the film had melted, resembling lukewarm beeswax.[83] Reed ordered the men to continue to examine the bag. During the trip across the Atlantic, mildew had grown, covering the entire surface of the envelope. The aluminum powdered paint on the silk and cotton bag was flaking off in large chunks.[84] The cement between the seams of the fabric had deteriorated, leaving large gaps between each panel. Reed felt disheartened and reported the findings to Major Thornell, who immediately issued a request for a new envelope. Once again, the War Department denied the request, sending a reply that insinuated patching the envelope would work well enough for the time being.[85] The tedious job of patching each tear, cementing every seam and removing all of the mildew was underway. The men worked tirelessly day and night to prepare the envelope for *ROMA*'s first hydrogen test.

Workers fit the rigid nose copula to *ROMA* during her reconstruction at Langley Field. *Harry Dale Treadway, courtesy of Richard Treadway.*

On August 31, news broke that several blimps and balloons had been destroyed in a fire at the navy's Rockaway Air Station in New York, including the navy's largest non-rigid ship, *D-6*. The fire was caused when the spark of a workman's hammer ignited the hydrogen, lighting off the gas tanks of the airship. Because there were no casualties, this was also dismissed as a mere incident. The War Department decided that extra precautions would be taken at airship hangars to avoid a similar situation in the future.[86]

As the crew arrived to *ROMA*'s hangar the morning of September 14, the soles of their shoes were checked for metal tacks and nails,[87] and all cigarettes and lighters were ordered to be left away from the hangar. This was the day that the crew would perform *ROMA*'s first hydrogen test in the United States.[88] Under the direction of the War Department, there would be no metal tools working in the hangar. The officers emphasized the combustible nature of hydrogen, pointing to the large signs that were prominently nailed to the outside of the hangar. Each sign read: "DANGER! NO SMOKING WITHIN 50 FEET OF THIS HANGAR BY ORDER OF C. OF A.S."[89]

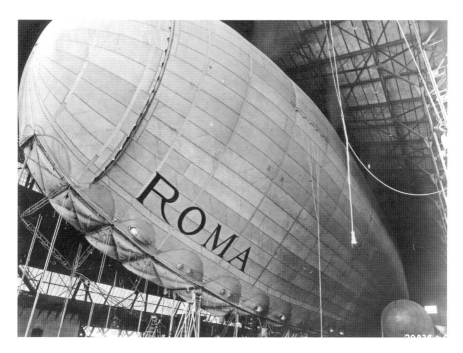

ROMA prior to the fitting of her nose copula as her articulated keel was being reassembled. *Hampton History Museum.*

Alberto Flores and Virgil Hoffman jumped into action as they and the other riggers tied *ROMA*'s envelope to pulleys attached to the concrete floor inside the hangar. Other crew members placed hoses connected to hydrogen tanks inside the bag. At 9:40 a.m., the gray skin of *ROMA*'s envelope shook and shuttered with a cataclysm of waves as the gas flooded into the ballonets. From bow to stern, *ROMA* grew in size, no longer resembling a large withered silver patchwork sheet. The wrinkles evened as the fabric grew increasingly taut, causing a surge of excitement among the crew. The men cheered when her envelope was finally full. The floating ship filled the hollow recesses of the hangar. *ROMA* had come to life.

The autumnal leaves began to turn hues of red and gold as September became October. The reconstruction of *ROMA* continued, with her men working day and night. The War Department was putting pressure on the air service to get *ROMA* in the air as soon as possible in order to absolve the hydrogen-fueled fears of the public and sway them back to embracing lighter-than-air technology.

The lattice copula was assembled outside the hangar. Men climbed on ladders along the frame, carefully putting each piece together, like spiders

ROMA during the reconstruction of her articulated keel and engine mounting at Langley Field. *Harry Dale Treadway, courtesy of Richard Treadway.*

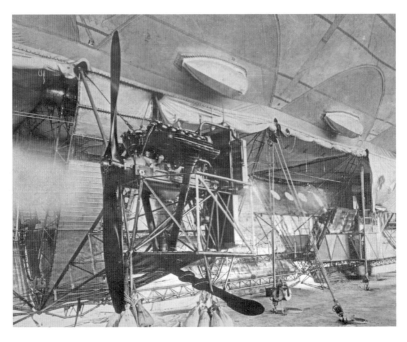

A view of one of *ROMA*'s Ansaldo motors, as well as her partially completed keel and cabin. *Air Combat Command History Office, Joint Base Langley-Eustis.*

Master Sergeant Harry Chapman stands in front of *ROMA*'s partially assembled nose copula. *Hampton History Museum 1987-18-23.*

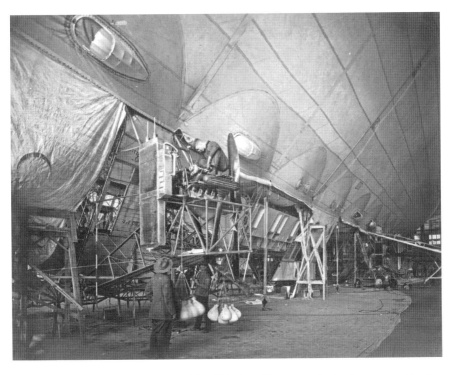

ROMA crew members work on one of the ship's Ansaldo motors during her reassembly. *Air Combat Command History Office, Joint Base Langley-Eustis.*

carefully constructing a web. The triangular keel was tied along the bottom of the envelope, each piece of the articulated joints and hollow tubes carefully examined and put together like a large jigsaw puzzle. The blueprints provided by the Italians were constantly studied, and debates were often held over how each piece fit together. Assembling this ship was an intensive labor of love.

The U.S. Army Air Service's top brass began discussing how they would use *ROMA* in this era of peace. For the immediate future, she would be used to train new airmen entering the balloon companies. Eventually, she would be used to conduct experiments in reconnaissance and aerial photography. *ROMA* would be fitted with two new compartments to aid in this: a radio compartment and a state-of-the-art photography space.[90] She would have a powerful radio system that would allow the ship to have two-way communications not only with the ground but also with other aircrafts.[91]

The photography space would consist of a studio that would allow the ship to take aerial photography and process the film in an on board

ROMA, full of hydrogen and anchored to the ground, as her articulated keel and nose cone are reassembled. *National Museum of the U.S. Air Force.*

Members of *ROMA*'s crew sit on a small crate in front of the partially assembled airship. *Air Combat Command History Office, Joint Base Langley-Eustis.*

darkroom.[92] Within thirty minutes of taking an image, it could be processed and then dropped via parachute to the desired location.[93] Students from Langley's photography school would train on board *ROMA* for a future of aerial reconnaissance photography in the service. The eventual goal was that these new additions would make *ROMA* a key player in military aviation.

The buzz about *ROMA* continued to flood newspapers across the country. Readers salivated over headlines of the army's plans to tour the ship across the country. Newspaper rumors of the airship performing as a peacetime transport for civilian passengers made readers dream of seeing the country from the sky above, to be able to travel from Virginia to Ohio[94] or even Los Angeles[95] in what seemed like the blink of an eye. The spectrum of travel would be broadened, and for the first time, places once inaccessible seemed to be at their fingertips. *ZR-2* and the other airship accidents soon faded from headlines, replaced by the excitement and grandeur of *ROMA* and her crew.

ROMA in her hangar at Langley Field shortly after the fitting of her rigid nose copula. *Hampton History Museum 1987-18-22.*

(Back from left) Sergeant Hoffman, Master Sergeant McNally, Master Sergeant Chapman, Sergeant Biedenbach, Sergeant Beall, (front from left) Lieutenant Reed, Major Thornell and Captain Mabry. *Air Combat Command History Office, Joint Base Langley-Eustis.*

ROMA sits in her hangar at Langley Field after her reconstruction. *Hampton History Museum 1987-18-35.*

Leaves began to fall from the trees, and the breeze turned brisk as October faded into November. Final checks were made on the keel, box-kite rudder, controls, nose cone and the envelope.

On November 5, 1921, with great glee and relief, *ROMA* was finally complete and ready to take to the skies.[96]

CHAPTER 5

THE FIRST FLIGHT

"How do we look from the ground?"
"Magnificent!"
—*Major John G. Thornell to radio operator at Langley Field*[97]

News quickly traveled around Hampton Roads that, finally, *ROMA* was going to take to the skies. On the morning of November 15, over one thousand people traveled to Langley Field to watch American aviation history being made. Loved ones of the crew gathered in a cluster near the hangar, excitedly chatting about the celebrity status of their airmen.[98] Exhilaration was positively palpable in the sunlit air as the crowd waited for the military's largest aircraft to take to the American skies for the first time.

Captain Dale Mabry emerged from the hangar and waved to the crowd. Through a cardboard megaphone, he ordered the two-hundred-man handling crew to pull *ROMA* from her metal abode.[99] The men edged her out slowly, clinging to the ropes mooring her to the ground.[100] Cameras rolled, fiercely filming for newsreels, and reporters from various news sources ran about, collecting as much as they could about the scene. With a collective gasp from the audience, *ROMA* emerged, her enormous silver bag glistening in the vibrant morning light. The unknowing crowd was blind to the almost two hundred patches covering her feeble envelope.[101] The air crew appeared, keeping a steady watch on their ship.

Curious children tugged at their mothers' skirt hems with questions about the mighty ship while men nodded their heads approvingly at this engineering

ROMA emerging from her hangar at Langley Field. *Air Combat Command History Office, Joint Base Langley-Eustis.*

marvel. Crew members, positively beaming in the shadow of *ROMA*, waved to their sweethearts, loved ones and adoring admirers. Engineers boarded their engine platforms, and with a brilliant roar, *ROMA* was given life.

Lieutenant Reed caught a glimpse of his wife and smiled at her. With a deep breath, she returned his affectionate gesture. Turning to the other wives and family members, she regaled them with stories of their flight in Italy, assuring the more nervous ladies of *ROMA*'s safety. A few of the women looked on this boisterous woman with envy, but most clung on to every word she said.

Stella Hoover stood nearby, craning her neck to catch a glimpse of Virgil Hoffman. She shivered in the cold and choked back a lump in her throat. She was nervous for Virgil's safety after all of the recent news. He reassured her that *ROMA* was a safe, sturdy ship and that she had nothing to worry about. But the young bride-to-be wouldn't feel better until her Virgil was safely back on the ground.

Major Thornell was anxious to prove *ROMA*'s worth to the War Department. That day, Colonel Charles Danforth, commanding officer of

Stella Hoover and Sergeant Virgil Hoffman sit together at the beach as Hoffman smokes a cigarette. *Hampton History Museum 1987-18-128.*

Langley Field, and Lieutenant Colonel Albert Fisher, chief of Langley's lighter-than-air operations, would be on board as official observers.[102] Corporal Alberto Flores climbed through the envelope to his position in the crow's nest.[103] He stretched his neck out the door and examined the topside

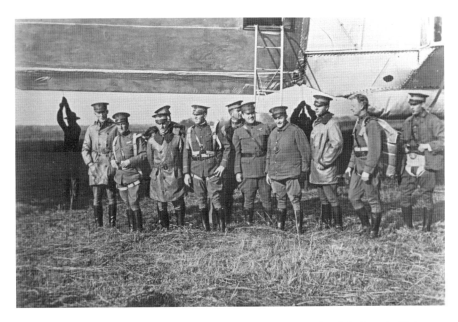

Officers from the balloon school and members of *ROMA*'s crew shortly before the flight on November 15, 1921. *Air Combat Command History Office, Joint Base Langley-Eustis.*

ROMA shortly before lifting off for her trial flight on November 15, 1921. *Air Combat Command History Office, Joint Base Langley-Eustis.*

ROMA lifts off from Langley Field during her November 15, 1921 test flight. *National Museum of the U.S. Air Force.*

of the bag. All looked tight and well, so he sent word to the control cabin. It was time for *ROMA* to fly.

With a signal from Major Thornell, the handling crew slowly let go of the thick ropes and *ROMA* ascended. Photographers snapped pictures, and newsreel cameras rolled to capture the monumental event. A rush of adrenaline filled the air as the spectators let out a loud cheer of excitement and applause. The crew was too busy to watch as the world rushed around them and the ground was farther from their reach. *ROMA* rose statically without issue.

Once Major Thornell felt the ship had reached proper altitude, he ordered for a full stop and rushed to the radio compartment. Messaging down to Langley Field, he asked, "How did we look when we left the ground?" There was a momentary silence and static. The crew held their breath.

The radio operator from Langley replied, "Magnificent!"[104]

Boisterous cheers and clapping echoed throughout the ship and the men patted each other on the back. Major Thornell grinned and, pleased, exclaimed, "Good job, boys! Good job!"

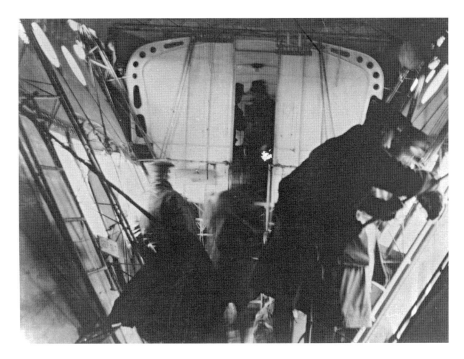

An interior view of *ROMA*'s passenger cabin while in flight. *Hampton History Museum 1987-18-37.*

ROMA stayed aloft above Langley for an hour while the crew performed a thorough preflight inspection. In the control cabin, Captain Mabry bustled about, performing his duties as second in command. Lieutenant Byron T. Burt Jr. was in position as the directional pilot for *ROMA* while Walter Reed was on the opposite side of the control cabin as the altitude pilot.[105] Master Sergeant Harry Chapman watched the pressure gauges anxiously, monitoring the purity of the eleven ballonets inside the bag. With the accidents of *C-3* and *ZR-2* not being far behind them and knowing the fragile state of the envelope, Harry felt a great deal of pressure to monitor the purity very closely. The gas compartments had to maintain at least 90 percent hydrogen purity by regulation of the air service.[106] To his tentative relief, Harry recorded that the purity levels were averaging above the standard.

Toward the end of the inspection, "Little Flores" burst through the door to the control cabin. Startled, all of the men looked up. He was clutching the arm of a young man whom no one recognized. Mr. Marleon was a local man who had stowed away in the copula, vying to be part of aviation history. Major Thornell showed mercy on the man, and since they had nowhere else

ROMA in flight with engineers on the platforms inspecting the damage caused by a door flying into a propeller. *Hampton History Museum 1987-18-45.*

to put him, ordered him to work on the ballast on the rear sand rack under the direction of Private Gus Kingston.[107]

"All right, men, let's show the people what she can do. Turn the engines to 1,100 RPM. Take her across the river!" ordered Major Thornell. At 10:30 a.m., the order was sent throughout the ship's telegraph system. Engineers attached themselves to harnesses and precariously walked on the scaffolding to their individual engines. The spectators on the ground held their breaths as they watched the engineers perform this daredevil feat, as if they were tightrope walkers at the circus.

At the left center motor, Jethro Beall discovered frost on the carburetors of his motor. The Ansaldo engines were designed to run oily, which worked well in the tepid southern Italian climate—not the wintry Virginia gusts.[108] After a great deal of chipping and prodding, he was finally able to bring his engine to life. The power wasn't nearly what he observed in Italy. He shook his head and retreated inside the ship. There he was met by Joe Biedenbach, who was in charge of the engine on the opposite side of the ship from Beall.

"I can't get the damn thing to go above 600 RPM!" Jethro exclaimed.

"Neither will mine. These engines weren't built for winter weather!"[109] Joe called back.

The ship flew at a pace of forty-four miles per hour away from Langley Field. People began coming out of their homes to watch the airship fly over. Men leaned out of office windows, and children gathered in their front yards. Thornell ordered a message relayed to Langley Field: "We are sailing very nicely and everything is well…The weather conditions are grand!"[110]

When passing over downtown Hampton, Walter Reed briefly switched positions with Dale Mabry. In his hand, he clutched a weighted canvas message bag. He slipped in a note that read: "To Maria, Everything is splendid…I will see you at dinner. I love you." He glanced out the window and saw the ship perfectly positioned over top of his in-laws' large, white house that sat along Hampton River. Walter leaned out the window and dropped the bag down, knowing it would land in the yard. The note would grant a sense of relief for his young wife.[111]

Girard Chambers Jr., a twelve-year-old boy working alongside his father at Hampton Normal and Agriculture Institute,[112] paused and watched the airship. He twisted his mouth into a scowl. "Isn't *ROMA* grand?" his father asked him.

The young boy scrunched his nose and thought for a moment. "That ship looks more like an elephant,"[113] he replied. His father laughed and patted him on the back.

From *ROMA*, the white lighthouse at Old Point Comfort was clearly visible next to the moat that encircled the stone walls of Fortress Monroe. Braving the chilly wind that blew off the Chesapeake Bay, people gathered on the beach to watch the dirigible fly over. Patients lined the porches of the post hospital alongside nurses who chatted with quiet excitement. *ROMA* glistened in the morning sun as she smoothly sailed through the sky. Everything was going smoothly and splendidly as the ship crossed over Willoughby Bay. Crowds gathered along the shoreline of Willoughby Spit, and visitors at the Pine Beach Hotel sat on the front porch, gazing to the sky. At the navy base, sailors sat on the decks of the ships, pointing to the dirigible. At the nearby Army Quartermaster Depot, the men gathered along the maze of railroad tracks that ran throughout the post and silently watched as *ROMA* flew over. Photographers climbed to the tops of apartment buildings and warehouses to capture images of the great dirigible. At that moment, it was as though *ROMA* paused time and all other cares disappeared. She tugged at the imaginations of all the onlookers who dreamed of taking to the skies and seeing the world from above.

Shortly before 11:30 a.m., Sergeant Lee Harris sat in the doorway to the outrigger platform for the forward port engine. He was startled when a small aluminum door from the forward section of the keel broke free and crashed into his engine's propeller. Wooden chunks of the propeller blade flew in all directions, tearing the thin fabric covering the keel and along the bottom of gas compartment 3. Sergeant Harris immediately jumped into action. He rushed out onto the platform to shut the engine down. Then he hurried back into *ROMA*. There he was met by Master Sergeant William Fitch and Charles Brannigan. The three men knew that any leak in the gas compartments could be devastating. They had mere moments to avoid a similar fate of *C-3* and *ZR-2*. They climbed the ladder into the envelope until they reached a long tunnel that ran through each of the ballonets. After reaching gas compartment 3, they briefly choked on the hydrogen gas that began filling their lungs. Regardless of the circumstance, they went steadily to work. With cement and patches in hand, the three men worked diligently until they began to lose consciousness.[114] Their quick thinking and meticulous repairs were enough to maintain the compartment's gas purity, thus saving the ship and their crewmates.

At 1:00 p.m., *ROMA* returned over Langley Field. Major Thornell ordered all engines stopped, and the mechanics on board made their way along the platforms to their engines. The onboard riggers threw ropes out from the ship and Corporal Flores tossed one from his perch on the front of the ship. Three hundred men gathered on the ground, reaching to the sky to catch the ropes dangling from the dirigible. Gas valves on the ship were opened and *ROMA* began its slow descent.

Within a half hour, the ship's keel touched the ground once again and the crew disembarked. Sergeant Harris, Master Sergeant Fitch and Mr.

ROMA above Langley Field shortly before landing after sustaining damage in flight. *Hampton History Museum 1987-18-42.*

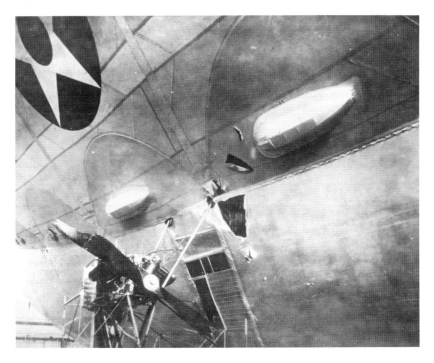

A close-up view of the damage caused by a door flying into and shattering a propeller. *Air Combat Command History Office, Joint Base Langley-Eustis.*

Brannigan were thoroughly examined by an awaiting medical team before being declared in satisfactory enough condition, especially given their harrowing feat. Reporters rushed the crew, thrusting a squall of questions toward the men. A reporter from Hampton's *Daily Press* approached Dale Mabry and asked how he felt the voyage that day had gone. Straightening his shoulders and smiling, Mabry replied, "The *ROMA* is a splendid addition to the American Air Service!"[115]

Films were hurried to nearby theaters. That evening, newsreels of the flight were shown throughout Hampton Roads. News media outlets prepared the footage to be shipped to other theaters across the country. Audiences around the nation would soon enthusiastically watch as this brilliant gem of the Air Service grandly ascended to the sky. The stowaway, Mr. Marleon, returned to his home that evening, boasting to his family and friends about his adventure on board *ROMA*. Major Thornell spent the evening writing his report to the army's brass. Before closing and signing his name, he made sure to add, "Sergeant Harris was in charge of the motor group at the time of the

Stern view of *ROMA* shortly before takeoff showing her triplane box kite rudder assembly. *Air Combat Command History Office, Joint Base Langley-Eustis.*

accident and he deserves special commendation for his good judgement and quick action. It is believed that this soldier should be mentioned in orders for his excellent conduct under very trying conditions. The other two men should receive letters of commendation."[116]

The next day, the *Daily Press* ran a story detailing the flight for its readers. The article described how *ROMA* moved "as gracefully and as pretty as the flying of the American eagle."[117] *ROMA* was no longer just a symbol of friendship between Italy and the United States; she was now a symbol of freedom.

CHAPTER 6

CHRISTENING AT BOLLING FIELD

We almost lost the ship several times…We had some trouble with the motor.
—Sergeant Virgil Hoffman, in a letter to his brother[118]

*R*OMA became the subject of publicity and was used as a key recruiting tool. Newsreels ran continuously in the local theaters. A recruiter from Langley Field was sent into town every day of the week, attempting to persuade young men to join the lighter-than-air division.[119] Walter McNair, a physicist from the Bureau of Standards,[120] arrived from Washington, D.C. He had come to Langley with the purpose of installing devices on board *ROMA* meant to measure the ship's air speed.

Patchwork continued daily on the deteriorating envelope and ballonets. From the accident that had occurred during the trial flight, a hole the size of a large fist was clearly visible.[121] The engineers continued to express their growing reservations over the Ansaldo engines. The general consensus was that it was necessary to replace them with the lighter, more familiar, American-made Liberty engines.

Jethro Beall found himself with increasing trepidation toward *ROMA*. He realized that she wasn't quite as strong as everyone had been led to believe. Watching what Sergeant Harris, Master Sergeant Fitch and Mr. Brannigan endured shook him very deeply. This wasn't what he had envisioned flying on this airship would be like. An ominous sense of dread spread throughout his body. To his old army buddy Zenos Uland, he wrote:

ROMA over the city of Norfolk, Virginia, during her successful November 23 test flight.
National Archives and Records Administration.

Zenos,

Yes, I am on THAT ROMA, and we are all working night and day to get it ready for flying again. This ROMA is nothing but a death trap! Three men were gassed today while working inside the bag. Someday it's going to wrap itself around a tree and then all will be over. Everybody in the service has the same idea about this ship.

Please write soon,
Jethro[122]

A second trial flight went forth on November 23,[123] this time without incident. Mr. McNair was able to test his new instruments and return to Washington, D.C., delivering high praise to the government about both *ROMA* and her crew.

After returning home with the remains of the victims of *ZR-2*, Lieutenant Clifford A. Tinker was released from service from the U.S. Navy. As a civilian, he was able to speak more freely about his fears regarding the continued use of hydrogen and turned to his former colleagues in the military, urging them to fill American airships with helium. Clifford Tinker emphasized the dangers of hydrogen, emphatically expressing that the higher cost to procure helium wasn't comparable to the lives that could be potentially lost because of hydrogen. Upon his insistence, the navy flew its blimp *C-7* in three trial flights between Norfolk and Washington, D.C., with a gas bag full of helium. These flights were considered a success for the future of lighter-than-air technology. Helium didn't expand or contract as quickly as hydrogen, and the crew noted that, remarkably, there was no loss of helium on any of the flights. This proved a long-term economic advantage over hydrogen. The popular aviation industry journal *Aerial Age Weekly* declared helium to be superior in both safety and economics to its combustible hydrogen counterpart and reported, "Summed up, it can be said use of helium as a gas for the inflation of airships has been demonstrated beyond a doubt."[124] Despite the success and the proven overall superiority of helium over hydrogen, the military helium plant in Fort Worth, Texas, was quietly closed indefinitely within days of the successful flights of *C-7*.[125]

Sailors fill the navy's *C-7* airship with the non-flammable gas helium. *J. Sargeant Memorial Collection, Norfolk Public Library.*

An elaborate christening ceremony was planned for *ROMA* on December 9. The airship would fly to Bolling Field near Washington, D.C. That chilly morning, the crew reported to the hangar at 4:45.[126] Unlike the previous two flights, there was no audience to cheer on the ship or the crew. Waves lapped in the nearby river as a brisk wind blew across Langley Field. The men's cheeks stung in the chill, and they rubbed their hands together to keep warm. The grounds crew pulled *ROMA* from the hangar, and the men climbed on board. As had become routine, they took their positions as the sun began to rise. Thornell ordered the engines started, but the engineers found their Ansaldos completely frozen. Meanwhile, dignitaries and top military officials from both Italy and the United States began to gather at Bolling Field, awaiting the scheduled arrival of *ROMA* at 8:30 a.m.

Thornell ordered boiling water poured over the engines' radiators and called for the exhausted crew to manually turn the heavy propeller blades. After arduous hours laboring over the Ansaldos, four of the six engines limped to life. Thornell glanced at the clock; it read 8:00 a.m.

ROMA's handling crew working with the massive airship. *Air Combat Command History Office, Joint Base Langley-Eustis.*

He fumbled nervously, rushing to try to get *ROMA* ready to leave for Washington, D.C.

At Bolling Field, the impatient crowd's enthusiasm waned. When no word from *ROMA* arrived by 9:00 a.m., Major Oscar Westover, director of the air service's air production unit, phoned Langley Field to inquire on the whereabouts of *ROMA* and her crew. When informed that she was still on the ground at Langley, Major Westover immediately cancelled the event and politely excused the guests. Once everyone had departed, the aviator boarded a plane and made his way to Langley.

Upon his arrival, a humiliated and infuriated Major Westover questioned all of the men in both *ROMA*'s handling and flight crews. It was ultimately decided that the Ansaldo engines were to blame. However, Thornell's lack of communication with Bolling over the status of the ship overshadowed the mechanical failures they had faced that morning. Major Westover intensely questioned Major Thornell while Thornell insisted that the ship would have made it on the four running engines, though he was less confident about whether she would have been able to make the return trip to Langley. An enraged Major Westover informed Major Thornell that his failure to report a cause of delay to Bolling would be noted in his report, making sure to emphasize the official and personal embarrassment it caused him and the government.[127]

The ceremony was rescheduled for December 21. This time, Major Thornell was determined that *ROMA* would reach Bolling on time that day. Despite heavy winds and frozen engines, *ROMA* lifted from Langley Field shortly before 6:30 a.m.

At Bolling Field, celebratory decoration was hung and the order was given that the uniform of the day for the handling crew would be the garrison drab uniform instead of their usual fatigues.[128] Shortly before 9:00 a.m., the official party began convening at Bolling Field once again. Secretary of War John Weeks was joined by several officials, including Italian ambassador Vittorio Rolando-Ricci; the Italian army's chief of staff, General Giuseppe Vaccari; and the ship's sponsor, Miss Fonrose Wainwright (who was the daughter of the assistant secretary of war). Major General Mason Patrick, the chief of the air service, was joined by Major Westover and Major Percy E. Van Nostrand, who was in charge of lighter-than-air development for the air service.

After being enticed by Major General Billy Mitchell to see the great airship, Clifford Tinker was also in attendance that day. Having forged a new career for himself in writing for various publications across the country, Mr. Tinker was seen by Billy Mitchell as a way to tout the necessity of big airships for the army to the press. However, Mr. Tinker was only in attendance to impress on the brass the importance of using helium in the country's largest airship.

The ceremony was to be one full of pomp and circumstance to welcome the army's "queen," Miss Wainwright, who would travel in a free balloon that was decorated in elegant patriotic bunting[129] and break a champagne bottle of liquid air (or air cooled to a temperature condensing it into a pale blue liquid and stored in a vacuum-sealed container) against *ROMA*'s keel. Afterward, the official party was promised a ride aboard *ROMA*. They planned to fly over Washington, D.C., and Baltimore and then back to Bolling Field.[130] Major Van Nostrand invited Mr. Tinker to participate in the exhibition flight. Mr. Tinker readily quipped, "I would gladly take a trip on her if they would fill her with helium."[131]

Major Van Nostrand looked very seriously at Mr. Tinker and quietly replied, "There isn't money to transport the helium from Texas to Langley."

Flipping through a notepad, Mr. Tinker retorted, "Well, I've calculated that you could transport the helium from Fort Worth to Port Arthur by train and then load it onto a navy ship to transport to Langley. I figured that, including shipping and handling, would be around $14,000."

Major Van Nostrand diverted his gaze and mumbled back, "We don't have $14,000 for transporting supplies."[132] Major Van Nostrand returned to the other guests while Mr. Tinker retrieved a pencil from his coat pocket to take note of the conversation.

Everyone gazed to the sky, waiting to see the great ship. As the minutes passed, the small crowd began to shift their feet and clutch themselves to keep warm. The wind gusts were merciless on the guests waiting in the open field. The Army Band, which played on a platform decorated with both American and Italian flags, would occasionally break through the squalls, playing music not only to distract the guests but also to keep their instruments warm.[133] Notes were frequently brought to Major General Patrick or Major Westover with radio messages from *ROMA*, stating that the ship was still en route. In an effort to avoid further embarrassment, Major General Patrick ordered a few planes to fly from Bolling in an attempt to ascertain the whereabouts of *ROMA*.[134]

Shortly before noon, a member of the handling crew pointed to the sky and yelled, "There she is!" *ROMA* was violently shuddering from side to side, appearing uncontrolled in the fierce wind. The fabric of her bag rippled and rumbled; the newly repaired aluminum door in the forward part of the keel whipped violently open and closed.[135] The guests watched motionlessly as the crew members tried to bring their helpless ship safely to the ground.

The box-kite rudder swung aggressively, creating incredible strain on the articulated keel.[136] A halyard holding the American flag to the stern had broken, and the flag was hanging upside down. The guests couldn't help but notice the irony as this was a universal sign of distress.[137] It wasn't until the ship drew closer that it became apparent that two of the six engines weren't running.[138] Ropes were thrown from the ship's cabins, and the handling crew members scurried to grab them. They frantically tried to pull *ROMA* from the sky. Dale Mabry leaned from the control cabin window, yelling down to the handling crew in an attempt to coordinate their efforts. The handling crew ran about beneath the airship, pulling with all of their might, but the untamed wild fury of the wind kept whipping the ship upward. It took a great many men on each rope, using whatever strength they had, to pull, edging *ROMA* closer and closer to the ground. With great concentration, exertion and strength, *ROMA* finally touched down. The handling crew grabbed on to whatever they could to hold the ship still as the gusting wind wafted against them.

Above and opposite, top: *ROMA* fights gusty winds while coming in for a landing at Bolling Field. *Hampton History Museum 1987-18-34 (above). Air Combat Command History Office, Joint Base Langley-Eustis (opposite, top).*

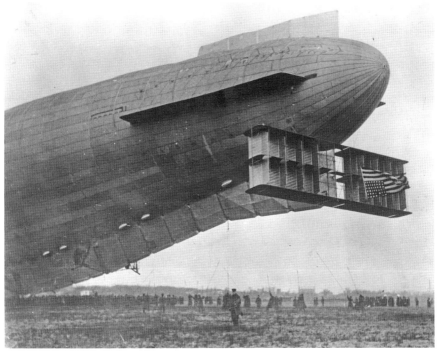

ROMA lands at Bolling Field with her American flag hanging upside down. *Hampton History Museum 1987-18-32.*

The weary flight crew disembarked, heaving a collective sigh of a relief. The officers shook hands with the guests, and John Thornell reported to Major General Patrick and Major Westover on the events of the flight. Three of the motors had given out, and John went on to praise the talent and expertise of his engineers for having been able to restart one of the engines in flight.[139] He shared the profound concerns of his crew members over the Ansaldos and expressed their overwhelming desire to see them replaced with engines more attuned to their training and the climate. Major General Patrick listened intently, nodding in agreement. He announced that he was canceling the exhibition flight, ordering *ROMA* back to Langley upon the conclusion of the ceremonies.[140] The gale-force winds were too strong and unpredictable to risk

Ms. Fonrose Wainwright, *ROMA*'s sponsor, waits as Ambassador Roland-Ricci delivers remarks at *ROMA*'s christening ceremony. *Library of Congress.*

Opposite, top: Keel view of *ROMA* shortly before takeoff. *Hampton History Museum 1987-18-61.*

Opposite, bottom: Members of the *ROMA*'s ground crew hold the ropes that are holding the airship to the ground after the ship landed at Bolling Field. *Hampton History Museum 1987-18-54.*

anyone's safety. It was also decided that for Miss Wainwright's safety, her free balloon would be replaced by a simple stepladder.[141]

Ambassador Rolando-Ricci stood in front of the crowd. He was a large man with a jovial smile, bushy beard and equally impressive moustache. He was obviously not accustomed to the wintry temperatures of Washington, D.C., and was wearing an ill-fitting fur-lined overcoat, accented by a boutonniere in the lapel. He wasn't wearing gloves, and his hands were shaking in the frigid temperatures.

Secretary Weeks, a man whose face closely resembled that of an English bulldog with a coarse moustache, joined the ambassador. He carefully removed his fedora, holding it tentatively in his hand. Behind the two men, *ROMA*'s crew members huddled in clusters together, quietly chatting.[142] What would have normally been interpreted as rudeness was largely disregarded due to the cold and hurried nature of the return of the ship to Langley.

With grand hand gestures, the verbose Ambassador Rolando-Ricci praised the War Department for keeping the ship's original Italian name. Turning to Secretary Weeks, the ambassador said, "I thank you for this courtesy, which is a courtesy to Italy. I consider its retention a fine omen for your aeronautics, inasmuch as nothing so greater than that which is called *ROMA!*"[143] After thinking for a moment, he took note of the day's proximity to the Christmas holiday. After a momentary pause, the ambassador concluded his speech by adding, "This Italian dirigible will never carry death and destruction but rather may descend from it to the earth the promise which from the sky of Bethlehem the angels made—Peace on earth to men of good will!"[144]

Secretary Weeks returned the gesture by officially accepting the ship on behalf of the War Department. He complimented the Italian engineers and designers for their talent and the ingenuity.[145] At the end of his speech, Secretary Weeks made a point to thank Major Thornell and *ROMA*'s crew for their efforts in arriving to Bolling Field, despite the adverse conditions they faced.[146] The guests clapped politely, though they were barely able to feel their hands in the sting of the wintry day. Miss Wainwright was presented with a large bouquet of roses as a gesture of thanks for her role as the dirigible's sponsor.

General Vaccari joined Major Thornell in front of the crowd. He sported the uniform of an Italian army infantryman with a grayish blue pillbox hat and a double-breasted coat covering his uniform.[147] His great, bushy handlebar moustache only served to accentuate his particularly serious face. General Vaccari handed Major Thornell a new American flag, and polite words of friendship between the two militaries were exchanged. The battered flag was removed from *ROMA*'s stern and the halyard fixed. The new flag was hoisted on *ROMA* as a cheer rose from the guests.

The crowd moved hesitantly beneath the bow of the airship, and a rickety wooden ladder was placed underneath the keel. John Thornell, Joe Biedenbach and a few of the other men steadied it as Miss Wainwright looked at the ladder skeptically. After receiving reassurances from Majors Westover and Van Nostrand, she handed her bouquet of roses off to another guest. She carefully climbed the ladder, carrying a large champagne bottle of liquid air in her hand. Major Thornell reached up his hand to aid her, which she readily accepted.[148] She tried desperately to maintain her modesty as her skirt rippled violently in the wind against her dark stockings. Lifting the large bottle above her head, she swung and smashed it against the keel. The glass shattered, and a cloud of bluish-gray smoke filled the air. Without waiting for applause or thought of her own grace, Miss Wainwright rushed off the ladder. With great relief washing over her face, she was handed back her roses, and she retreated

ROMA safely on the ground at Bolling Field for her christening ceremony. *Hampton History Museum 1987-18-33.*

into the crowd. After final words were given, the guests dispersed, and the crew began to prepare for a return to Langley.

Shortly before 1:00 p.m., the wind became more turbulent, and *ROMA* rolled onto her left side.[149] Panic-stricken, the crew desperately pulled her upright. Upon inspection, Major Thornell discovered that several pieces of one of the nonworking motors had been crushed. Along with other far less pressing damage, there were tears in the fabric that encased the keel section.[150] Major General Patrick ordered several crew members to return to Langley by train.[151] What could be repaired at Bolling Field hastily was, and the ship took on water and fuel.

At 2:30 p.m., *ROMA* took to the sky to head back to Langley Field. Virgil Hoffman stood on the ground, watching *ROMA* speed away into the sky. Virgil and several of his crewmates made their way to the train station for a far less harrowing journey home than experienced by those still on board *ROMA*. The battered ship limped into Langley at 6:30 p.m.[152] with only one working engine.[153] The envelope looked beaten and tired. The crew disembarked, and *ROMA* was returned to her hangar. John Thornell granted his exhausted crew a few days liberty to celebrate the holidays and rest.[154]

That night, Virgil penned a letter to his brother, Earl, about the journey to Bolling Field that day: "[T]here was some wind. We almost lost the ship several times but she was pulled into the hangar and put to rest for a while…We had some trouble with the motor."[155]

Chapter 7

NEW ENGINES AND "THE GRAVEYARD CLUB"

Every day lost in the use of the ROMA *in actual flight is a handicap upon lighter-than-air airship development.*
—*U.S. Army Air Service, December 30, 1921*[156]

Even the strongest of the crew's spirits was shaken by the voyages to and from Bolling Field. The winter storm that had plagued the flights on December 21 continued on for several weeks. Without a car and with the streetcars not running, Virgil Hoffman was unable to spend the reprieve with his dear Stella. He frequently wrote letters home to Michigan, telling his family all about her and sending pictures of the two together.

Lieutenant Burt enjoyed the respite from duty to spend with his wife, Willie, and their son, who would often emulate his father, donning a miniature version of his father's military uniform.[157] When *ROMA*'s supply officer, Lieutenant Ambrose Clinton, received temporary orders to the airship school at Langley, Byron Burt volunteered for the collateral duty.[158] Walt and Maria Reed were often found entertaining their circle of friends, celebrating his recent promotion to the rank of captain.

The christening ceremony would be the last time *ROMA* would fly under Major John Thornell's command. He had received orders to transfer to Washington, D.C., and on December 31, Captain Mabry was named the commanding officer of *ROMA*.[159] It was a bittersweet moment, though Thornell was partially relieved to hand the reins over to the dashing young captain.

Sergeant Joseph Biedenbach, back center, and other unidentified members of
ROMA's crew. *Hampton History Museum 1987-18-138.*

As one of his final duties as commanding officer, John Thornell petitioned the U.S. Army Air Service to replace the Ansaldo engines. He noted, "They were much too cold…and it was impossible to warm them up to running temperature."[160]

After witnessing the incidents in December, the air service readily approved his request. A letter was soon drafted to McCook Field in Dayton, Ohio, stating: "It has been decided that *ROMA* will not make another trip until it is equipped with Liberty motors."[161]

Engineers at McCook Field went to work, refitting six standard twelve-cylinder Liberty heavier-than-air engines for lighter-than-air use, including an enlarged propeller from the original engine design.[162] On January 7, 1922, under the direction of power plant manager Walter W. Stryker, a group of the best civilian engineers in the country left Dayton to journey to Langley Field with new engines in tow.[163] This was delightful news for many of the engineers assigned to *ROMA*, like Joe Biedenbach. After the perilous trip to and from Bolling Field, he wrote to his family that he felt that the Ansaldo engines weren't reliable.[164] Byron Burt was appointed the ship's engineering officer and was put in charge of giving the civilians help in whatever capacity he could.[165]

There would be very little to no training for the military engineers on the Liberties. The army had made it quite clear that it wanted to get *ROMA* back in the air as soon as possible.[166] Since these engines were the most widely used in the air service, the engineers were considered already competent on their design, functionality and usage. However, no one really had any idea how these engines would function on this semi-rigid airship design since all previous experience had been limited to either small non-rigid airships or, for the navy, the very large rigid zeppelins. The brass made a point to have the new Liberty engines tested before their arrival in Virginia, reducing the time *ROMA* wasn't flying. Almost as soon as the new engines arrived, the Ansaldos were unceremoniously removed from *ROMA*'s engine scaffoldings and sent to storage for posterity.[167]

The men who weren't assigned to aid the civilians spent their days cleaning cabins, patching and repairing the bag, inspecting the rigid framework or in temporary duty at the airship school.[168] There was a great deal of distress among the crew over the condition of *ROMA*'s envelope. The thin fabric of the envelope and ballonets was rapidly rotting away, and patch work continued daily. On behalf of himself and his crew, Dale Mabry wrote to the air service, requesting that the envelope be replaced as soon as possible:

February 1, 1922

Officer in Charge of Airship "ROMA"
Commandant, Airship School.
Envelope of Airship "ROMA"

The envelope of the airship "ROMA" is fast deteriorating.
Every man available has been at work on repairing the envelope, repair work having first started in October, 1921.
Practically every bias in the fabric over the air ballonets has been re-cemented, all bias in compartments 1 and 11 that have not been re-cemented or taped, are leaking. One hundred and eleven patches were made on the top of the envelope yesterday.

DALE MABRY,
Captain, Air Service[169]

After writing a considerable number of letters back and forth with the various departments of government able to subsidize such a purchase, the War Department decided to finally go forth and allot the funds to procure the new envelope. However, because it would have to be made in Italy and out of fine materials, it would take upward of six months to receive it. While nursing along a weakening air bag wasn't unusual, the considerable and rapid deterioration of not only the envelope but also the ballonets made it that much more treacherous to fly *ROMA*.[170] However, the War Department would not ground *ROMA* until the new envelope arrived.

ON FEBRUARY 9, 1922, TOP BRASS from the U.S. Army Air Service, including Major General Mason Patrick and Major Percy Van Nostrand went to Congress to meet with the Committee on Appropriations to discuss the budget of the army for fiscal year 1923. A particular topic of conversation kept turning to whether or not to fill the *ROMA* with the non-inflammable gas helium.[171]

"What did you say is the capacity of the *ROMA*?" asked the Honorable Daniel R. Anthony Jr., chairman of the committee.

"1,200,000 cubic feet," replied Van Nostrand.

"So, it would require nearly all of your entire amount of helium to give *ROMA* one filling?" asked Chairman Anthony.

"No, it would take about one-half of it."

Chairman Anthony thought for a moment. "Do you intend to use up the supply of 2,400,000 cubic feet you said you have on hand, or do you intend to keep it in storage?"

Major Van Nostrand cleared his throat and glanced over to Major General Patrick. "That will be a matter of policy to be decided, sir."[172]

The airship accidents that had occurred in 1921 became distant memories, and with the closure of the helium plant in Fort Worth, discussion over its immediate necessity to replace hydrogen was shelved.

Upon hearing the news, Clifford Tinker called on Representative Frederick R. Hicks, the chairman of the subcommittee that oversaw military aviation.[173] Mr. Tinker had come to know Representative Hicks quite well while serving as a publicity officer for the U.S. Navy just a few months prior. "Sir, you need to supply the funding to transport helium from Fort Worth to Langley for *ROMA*," Mr. Tinker emphatically pleaded.[174]

Representative Hicks sighed, "And why so?"

"After *ZR-2*, the whole country is alarmed over further accidents in airships; they are losing faith in airships. If an incident were to occur, this could damage the lighter-than-air programs of both the army and the navy," Mr. Tinker replied.

Representative Hicks sighed and rubbed his temples with his fingers. After a momentary pause, he waved his hand in the air and said, "Nothing doing."

Mr. Tinker was taken aback. "Nothing doing?"

"Mr. Tinker, Congress simply isn't interested in the lighter-than-air programs. Why would we expend more money on this program?" Representative Hicks replied, sounding quite exasperated.

"We're talking about the lives of every crewman on board *ROMA*! Can't there be some sort of remedy?" Mr. Tinker exclaimed.

"None that I know of unless they can get some funds that they can juggle," Representative Hicks replied.

"Sir, you cannot put *ROMA* into the air using hydrogen!" Mr. Tinker exclaimed.

"Mr. Tinker, the Army Air Service feels that every day lost in the use of the *ROMA* in actual flight is a handicap upon lighter-than-air airship development," Representative Hicks said. He diverted his gaze down toward his desk and busied himself with some paperwork.

"But, sir—" Mr. Tinker started.

Representative Hicks held up his hand to silence Mr. Tinker. "The *ROMA* will continue to fly with hydrogen. Now, good day, Mr. Tinker." [175] Clifford Tinker shook his head and left feeling defeated and helpless. The images of the bodies of the men who died on board *ZR-2* were quite vivid in his mind.

At Langley Field, the men exhibited a wide array of emotions regarding *ROMA*. Rather tongue in cheek, the officers nicknamed *ROMA* the "Suicide Ship." They felt that this ship was so strong and had endured so much that there was absolutely no way anything could ever happen to her.[176] However, to the enlisted crewmen and their civilian colleagues, this sentiment wasn't shared.

Robert Hanson, a civilian mechanic from McCook Field, penned a letter to his aunt in Ohio: "I might come home in this airship but let me tell you I am not very anxious to go that far. The fabric is in bad shape and they can't get a bag from Italy until about July or August. But they are going to fly it just the same, and we mechanics have to take three trips before we go, so I hope we have good luck."[177]

Several of the crew members shared the same concerns as Mr. Hanson. A few members of the crew formed what they nicknamed "The Graveyard Club,"[178] made up of men feeling a sense of dread toward *ROMA*. Jethro Beall was the self-appointed leader of the motley crew. His anxiety had boiled over into a terror that he was unable to harness or rationalize. Jethro petitioned for reassignment off *ROMA*, but he was continually denied. His own resounding sense of dread wasn't enough to transfer one of the best airship engineers from the premier dirigible of the air service. He penned a letter to his friend Leona Bell with what he felt would be a last testament.

Dear Leona,

The ROMA *and I are on the outs again for now that everything is ready to fly again I am trying to get off it. We have formed a new club of crew of the* ROMA, *with the name of "The Graveyard Club." What do you think of that for a name? It sure does fit the ship, as it is a death trap for sure. If it was to come down, no one would be able to get clear of it. This ship is a death trap for sure.*

Take care,
Jethro[179]

Despite mixed feelings regarding the safety of *ROMA*, work continued day and night to prepare her for flight. The ship seemed to present as many challenges in her maintenance as she did prospects for her future. Many defects in *ROMA*'s design and stability were becoming apparent, including warped doors, damaged engine catwalks and issues with the elevator controls.[180] Her gas tanks were discovered to be badly rusted and leaking, requiring constant maintenance.[181] But most pressing of all of the issues continued to be the condition of the envelope of the ship. The hangar log read like a sewing pattern of constant patchwork. It was as if the moment one leak was patched or cemented, another would be found. This was most pressing in gas compartments 1 and 11, where it was difficult to maintain the hydrogen purity due to the deterioration of the ballonets and envelope. One man was chosen each day to check the bag for leaks, often daring the impossible and hovering above the envelope most precariously, inspecting every millimeter he could. It was a most tedious task but nonetheless incredibly important. As each section passed inspection, a coat of aluminum paint was put on the bag to seal and protect the fabric. The air filled with the astringent scent of the banana oil used to mix the ground aluminum to create the paint.[182]

With the days and weeks waning and the detectible defects of *ROMA* temporarily resolved, the focus became ensuring the engines were properly installed and ready to go. The new engines were hastened off the testing blocks at Langley, giving the crew only minimal time to study them and make any adjustments needed to compensate for the weight difference between the Liberties and Ansaldos.[183]

A flight to break in the new Liberties had been scheduled for February 21, 1922. The civilian mechanics felt the pressure to finish in time under the direction of the lead civilian mechanic, Charles Dworack. He was an older man with the energy of someone far younger than his years, and his pinched face and square jawline were somewhat intimidating. He was an incredibly capable mechanic, perhaps the best the government had to offer. He had both a strong resolve and a serious air about him.

Meanwhile, Walter Reed found himself taken by the flu. After visiting the post medic, he was placed "sick in quarters" to recover. It was very uncertain whether he would make the flight on February 21. This was most distressing to Walt because aviators were only paid flight pay for their time in the air.[184]

Work on the engines finished two weeks ahead of the scheduled flight.[185] The remaining time was spent carefully inspecting *ROMA*, including the hull and copula, and a final layer of aluminum paint was put on the envelope. Dale Mabry was determined that this was going to be a flight of glory and would put to bed the blunders of the previous flights. It would also be a magnificent moment for him; everything he had trained for would culminate in this penultimate moment. Never one to shy away from the spotlight, Mabry was ready to put himself and his airship out in the public eye once more. He thought this flight would vanquish the apprehensions still left in the back of the public psyche regarding lighter-than-air travel and everyone would aptly see how dirigibles were the way of aviation's future.

News traveled quickly around Hampton Roads, and excitement built over the impending flight of *ROMA*. When the locals would hear the roar of the ship's engines, they would stop whatever they were doing to watch her overhead, lost in the imagination, innovation and excitement that *ROMA* had come to symbolize.[186]

With the winter storm finally passing, Virgil Hoffman was able to catch a break and visit his beloved Stella. The two sat together, fondly holding hands, smiling and looking at each other in the way that only two young lovers can. He shared with her the news of the planned flight on February 21. Stella expressed her worries, but Virgil was quick to reassure his bride-to-be that he would be safe and nothing would go wrong. Seeing the concerned look in Stella's large eyes, Virgil diverted the conversation, regaling her with stories of his family back home—of his brother, Earl; his sister-in-law, Nellie; and what a dear she was to his mother. He told her amusing stories of Earl and Nellie's children, who were always keeping Nellie busy washing and ironing the clothes they soiled and the trouble they were getting themselves into.[187] He told Stella more about his home of Eaton Rapids, Michigan, describing the picturesque small town in such great detail that Stella felt as if she were walking down the bustling main street of the tiny town. Virgil described the brick-front shops and the quintessential small train station where the coal dust from the train funnels would flitter in the air and stain the white-and-green paint on the building. Stella listened intently while playing with Virgil's silver cigarette case in her hands. She hung on to every word he said, feeling as though Virgil's family were her own, despite having yet to make their acquaintance. The two planned a trip together to Michigan once they were married.

When Virgil left that night to return to Langley, it felt as if someone had ripped Stella's heart from her chest. She knew the incredible ache would heal once she and her love were together again. This powerful bond was one that Stella thought only existed in the fairy tales of her childhood, but here it was, and she relished every moment she had with this man who so completed her heart.

Before leaving for Washington, D.C., John Thornell petitioned the U.S. Army Air Service to delay his orders and allow him to fly on board *ROMA* one last time.[188] This time, though, would be different; he would not be her commanding officer but merely an observer. This was a watershed moment for *ROMA*, and he felt he would be remiss not to be there to be part of such an important part of aviation history. The air service readily agreed to the postponement, and Marie Thornell went ahead to their new home without her husband. Fleeting separations were commonplace between the airman and his bride. There would only be a small amount of time before they would reunite once more. Without a tear shed, the two said a happy temporary farewell. Major Thornell took the opportunity to express to Dale Mabry how he wouldn't stand in his way while emphasizing that he would be there just as an observer. He knew that the crew had developed a deep admiration for the vivacious young captain and even the most pessimistic of the crew clung to Captain Mabry's zeal like a life raft for their own emotional and mental preservation.

The evening of February 20 was filled with merriment and fellowship for the officers of *ROMA*. They gathered at Langley's officers' club for a night of regalement and celebration of the events the following day.[189] Each officer's uniform was freshly laundered and pressed. Except for the occasional moustache, their faces were clean shaven. Their coifs were neatly slicked back in the style of the day. Caps sat starchily upright on their heads until they reached the entryway where, in proper military fashion, their covers were keenly removed.[190]

Each man was handed a small souvenir program, just large enough to fit inside the pocket of their jackets. The programs were decorated in an ornate design encircling an image of *ROMA* being drawn from the hangar, and a list of the crew was outlined inside.

The light in the room was dim, but the atmosphere was boisterous with laughter and conversation. Smoke from the men's evening cigars hung heavy and filled the air with the earthy aroma of tobacco. This was an opportunity to break away from the intense pressure that maintaining a fragile ship like *ROMA* required and enjoy the camaraderie of their brothers. These were bonds forged in the air during the Great War, built through months of training and flight, with an enduring resilience and trust in one another that couldn't be matched. In many ways, this crew was a family, with Dale Mabry as their patriarch.

Men filtered in and out of the club, with some coming in to escape the damp chill of the February night air and others escaping the stuffiness of the crowded room. Friends who were not slated to fly the next day arrived and reminisced over times past. They shared war stories and told tall tales of intrepid acts of bravery, raising their fallen brothers to monolithic proportions. Toasts were raised in their honor and in good fortune for the

The officers' club at Langley Field. *Air Combat Command History Office, Joint Base Langley-Eustis.*

days to come. A wry sense of morbid humor was often exchanged in light conversations. After surviving the aerial slaughter of the Great War and having chosen such a treacherous line of work, these aviators felt fortunate to be alive and to be able to make their undeniable mark on a new frontier of aviation for the U.S. Army Air Service. There could be no doubting that times were changing and that this was an age of optimism without caution, truly embracing a sense of carpe diem.

A young major by the name of John Reardon mingled in the crowd. Detailed to the Procurement Division of the U.S. Army Air Service and receiving instruction through the lighter-than-air school at Langley Field,[191] Reardon made himself somewhat of a protégé to John Thornell. He was a friendly man who had come to know all of the aviators at the field very well. He had come that night in hopes that he would be included on the crew manifest the next day. Like the rest of the men, he enjoyed the adventure that being in the air would bring. Anyway, he needed the flight time. Men fought for a spot on board the premier ship of the army's fleet. To be able to say that you flew on board *ROMA* meant that you were with the best balloon men that the U.S. Army Air Service had.

Those not regularly assigned to *ROMA*'s crew knew very little about the troubles she faced. What they heard was idle chatter, and the difficulties were often overshadowed by the premier position that *ROMA* held. Perhaps this could be seen as wide-eyed naïveté, the overconfidence of very competent airmen or the pressure from the top brass to make *ROMA* work, no matter her faults. After all, it would be an embarrassment to the army if she were to continue to have setbacks.

John Thornell and Dale Mabry found themselves at the center of all of the attention. Men flocked around them, falling over themselves to speak to the two commanding officers as if they were celebrities. Walter Reed was notably not present due to his illness. This was an odd moment for Mabry to not have his best friend and right-hand man present.[192] Nonetheless, he pressed on. Without hesitation, he took to the floor, readily answering the numerous questions thrown at him. Dale Mabry lived in this moment with the men he was charged with commanding over. John Thornell stood nearby, secretly relieved not having the burden of command on his shoulders or the inevitable backlash from superiors if something were to go wrong. Despite feeling a pinch of sentimentality, Thornell knew it was a new era for *ROMA* and that he was leaving her in good hands. He was looking forward to making one last flight on board before joining his wife in Washington, D.C.

Captain Frederick Durrschmidt approached Captain Mabry. A graduate of West Point, Durrschmidt had been nicknamed by his classmates "L'Allemand Duke,"[193] a reference to his German immigrant father and American mother. He had a sweet voice, soft eyes, carefully slicked hair and a small aviator's moustache. Durrschmidt looked more like the cinematic portrayal of an aviator rather than an actual birdman. He commented on Reed's illness and how it was unlikely he would be joining the flight the next day—no doubt, vying for a spot on the ship. Dale Mabry mentioned that Captain William Kepner would take Walt's place.[194]

Lieutenant William Riley stood by, listening intently to the conversation. He was a young man who exuded an intelligence justified by his Harvard education. "I see Burt isn't on the orders to fly," he commented. Lieutenant Byron Burt Jr. was scheduled to take some of the students from the airship school on board the small non-rigid airship *TC-2*, a ship that the men affectionately referred to as the "rubber cow." Mabry waved off the orders as he was determined to have Burt on *ROMA* the next day. He needed the best men on board the ship, and if there was anyone in the air service who understood the engineering of a ship, it was Burt.[195]

Captain George Watts, who was already included on *ROMA*'s crew manifest as a student observer, commented how he wished that the army would fill her with helium.[196] This sentiment fell flat among his shipmates. Another student observer, Lieutenant Ira Koenig, stood in passive thought over Captain Watts's comment.

After a few more hours of reminiscing, Captain Mabry clapped his hands loudly and said, "Let's break it up. We have a job to do tomorrow!"[197] A collective cheer rose among the men, and they dispersed to their respective corners of Langley Field for a good night's sleep before the early morning to follow.

For most of the men, it was a peaceful sleep unremarkable from any other. However, Ira Koenig was startled awake, sitting upright in his bed, covered in sweat. He had a nightmare in which he saw *ROMA* being chased through the sky by three smaller blimps. Suddenly, *ROMA* dove from the sky and crashed into the ground.[198] With his resolve shaken, he immediately sent word to his superior officer. Ira claimed that he had fallen ill and was unable to fly on board *ROMA*. Someone else would be able to take his place.

CHAPTER 8

FEBRUARY 21, 1922

If anything happens to me, you take me home.
—*Corporal Irbey Hevron to his friend Corporal Nathan Curro*[199]

The early morning of February 21, 1922, was hardly the "picnic weather" the crew had hoped for to test and break in *ROMA*'s new engines. Rain and sleet fell from the sky, beating heavily against the roof and walls of *ROMA*'s hangar with tremendous thumps and bangs. [200] The field surrounding the hangar was muddy and difficult to tread through. Despite the previous civilian turnouts to watch the ship, this cold morning, her fans either stayed in the warm surroundings of their homes, nestled beneath their warm blankets, or left early for work, hoping to catch a glimpse of the ship through a window or during a well-deserved break for lunch.[201]

Dale Mabry arrived at the hangar far earlier than necessary. His normally calm, cheery resolve was replaced that morning with stern anxiety. His face pinched in concentration, and his usual upbeat stride was more of a trudge. Mabry wasn't surprised to find Byron Burt at the hangar, inspecting the "rubber cow." Carefully looking over the instrument panel as a prospector would while panning for gold, Byron silently looked for a solution to his problem.

"Getting *TC-2* ready?" Mabry asked Burt.

"Sir, I think she won't fly today. It appears that she's broken," Burt replied, his face notably absent of concern.

Mabry cracked his famous smile for the first time since he had left the officers' club the night before. "I'll get you added to that roster."

"Aye, sir," Burt replied.[202]

THE NINETEENTH AIRSHIP COMPANY rhythmically marched in formation to the hangar. Once dismissed, the men trickled inside, shaking the rain off their uniform coats and grumbling over their mud-laden boots. The officers pulled at the collars of their uniform coats, the wet wool scratching against their skin. The hangar turned into a bustle of noise as the crew readied for the day ahead. Majors Thornell and Reardon soon arrived. Mabry asked if the two men would care to join him on his personal inspection of the ship, to which they readily agreed. They quickly found Master Sergeant Roger McNally and Sergeant Virgil Hoffman. Virgil had recently been appointed ROMA's chief rigger. They jumped into action to prepare ROMA for flight. Every patch and seam needed to be checked for leaks. The rudder and articulated sections of the keel were carefully looked over to ensure structural stability. Mabry was determined that this day would go without error, and if they had to fly in the rain, so be it. Then they could really prove to the War Department and Congress ROMA's worth.

Captain Mabry made his way into the control cabin, finding Little Flores along the way. He ordered the young corporal aft to watch the condition of the control surfaces of the rudder as he tested the controls. He first turned the directional wheel followed by the altitude wheel. After each spin, Little Flores reported back with positive results.[203]

Feeling confident in the controls, Mabry turned his attention to checking the gas cells. Harry Chapman stood near the glass manometer tubes, his eyes focused and lips pursed as he recorded into the ship's log the hydrogen purity within each ballonet. Without looking up from his work, he said to his commanding officer, "The purity is holding up fine, Captain. We are maintaining an average of 93 percent." Captain Mabry nodded approvingly.

Thornell turned to Mabry and commented, "You will be able to take additional students with that lift." Patting John Reardon on the shoulder, he continued, "Reardon, here's your chance to make the flight." Captain Mabry nodded in agreement, reminded of how so many of the students were salivating over being able to make the flight that day.

Like the focused bullet leaving the barrel of a gun, Mabry darted his way toward the ship's center catwalk en route to check on the engines, Majors Thornell and Reardon still at his heels. There they found Charles Dworack. "What have you done in the way of providing of the synchronization of the motors in flight?" Mabry asked.

"1,100 RPM is the determined cruising speed. When you signal 'cruising speed,' the starboard engineers of the fore and aft pairs will set at 1,100 and his opposite will then synchronize with them. On the center pair, the port engine will be the guide," Dworack replied.

Next, Captain Mabry found Master Sergeant James Murray, who was in charge of the ballast. Mabry asked after the status that day. "2,200 pounds of water and 5,250 pounds of sand aboard, sir," Master Sergeant Murray replied.

As Mabry walked away, John Reardon paused to inquire, "How is the ballast disposed?"

"The water is in two tanks located fore and aft; the sand is in a chute in the center and in bags in racks fore and aft, sir.[204] We're probably about sixty bags light on sand, though."[205]

"And these can all be disposed from the control cabin?"

"No, sir, there's no central switch to do so,"[206] Murray replied and then busied himself preparing for the impending flight.

John Reardon caught up with his inspection companions as they made their way back into the control cabin. Mabry was pleasantly surprised to find Walter Reed taking his post. Looking pale and with eyes a bit glazed over, he assured Mabry that he was feeling well enough to fly. That morning, he and Maria had a tense moment as she attempted to convince her husband to stay home and rest. However, she knew this was a fruitless endeavor. If *ROMA* was going up, so was her husband. Her concern over her husband's sickened state far outweighed her understanding of Walt's duty and the added incentive of flight pay.[207] After he left their house, she arranged to have a few of her friends over that afternoon for a game of bridge.[208] This would distract her for a while, and by the time they left, she would have just enough time to prepare dinner before he would return home from the hangar. Once this flight was over, they could argue about his need to recover some more.

Captain Mabry went to the hangar phone to call Captain Arthur Thomas, the airship school adjutant. He asked for five additional passengers to be added to the roster, including Byron Burt and John Reardon.[209] He also asked for the roster to be changed to reflect that Walter Reed would be on board in the place of Captain William Kepner.[210] Instead, Captain Kepner would be assigned to move another small blimp, *A-4*,[211] out of the hangar to make room to maneuver *ROMA* later that day.[212]

By 11:00 a.m., the sky began to clear and the preflight checks were complete. Captain Mabry dismissed everyone to grab a bite to eat and sent word to ensure that many of the civilian engineers would be joining them that day.[213] He was informed that Walter McNair from the Bureau of Standards would be on board that day to test a new air speed indicator.[214] Mabry fondly remembered the physicist from *ROMA*'s first trial flight in November.

Jethro Beall and Corporal Irbey Hevron met with a few of their buddies for lunch. It was obvious to everyone that Jethro was incredibly anxious about the flight. He remained mostly silent, his eyes distant as he pawed at his food, never taking a bite. Finally, Jethro said, "When we go up…we're not coming back."[215]

Everyone stopped for a moment to decide whether he was joking. When it soon became obvious that he wasn't, Irbey turned to his friend Corporal Nathan Curro[216] and said, "Well, if anything happens to me, you take me home."[217] The conversation turned to lighter matters, but Jethro continued to remain distant and stoic in his thoughts, like a man being led to his execution.

At 12:45 p.m., the crew returned to the hangar. That morning's rain and sleet had finally dissipated and the heavy mood the men carried with them turned to one of cautious optimism. They hoped that the change in weather was a good omen for the day ahead. Dale Mabry conducted another brief preflight inspection, and extra parachutes were added to the ship's storage. The new crew manifest was delivered to the hangar, and forty-five men were listed for the flight that day. They all began boarding and moving into their assigned positions.

As Major John Reardon climbed on board, he was greeted by fellow observer Captain Allan McFarland. With a cocky smile, McFarland asked, "So, are you a stowaway, sir?"

Reardon laughed and replied, "I'm just delighted to be a crew member in good standing!"

Lieutenant Riley approached the men from behind and commented, "Let's hope you'll still be delighted when the flight is finished, sir!"[218]

The passenger cabin soon became crowded with a large volume of student observers. The men helped strap one another into the bulky harnesses of the Stevens Pack parachutes,[219] grumbling over how none of them fit correctly. Major Walter Vautsmeier growled, "If this bag belonged to the navy, everyone would have his own fitted parachute."[220]

Captain George Watts chimed in with, "Yes, and we would be flying with helium."[221] Lieutenant Clarence Welch chuckled to himself as he helped Lieutenant Riley adjust his heavy parachute pack.[222]

In the control cabin, Byron Burt and Walter Reed were in position, checking their gauges and controls. Dale Mabry and John Thornell were also performing last-minute checks as Walter McNair fumbled with a gadget in his hand.

Alberto Flores poked his head through the crow's nest hatch. Once again he was in charge of watching the nose cone and the bag. Captain Mabry leaned his head out of the window and called up, "Little Flores, is everything all right topside?"

"Yes, sir!" he enthusiastically called back.[223]

Mabry then pulled his head in for moment, asking Burt and Reed if they were ready to take to the sky. "Aye, Captain," they each replied.

"Very good," Mabry replied. He then leaned his head out of the window and called to the handling crew chief, "Pull her out, boys!"

CHAPTER 9

ACROSS THE RIVER

She won't respond, Captain!
—Lieutenant Byron T. Burt Jr. to Captain Dale Mabry[224]

T he heft of *ROMA* required a substantial handling crew. A handful of the civilians from McCook Field not slated to fly that day volunteered to help. A gust of wind blew into the hangar, and the ship pitched violently. Little Flores, still perched out of the crow's nest hatch, yelled down to the handling crew, "Look! Hold her, fellas, or you'll knock me off of here!"[225]

One of the civilians on the ground, John Tucker, ran in front of the ship. He grabbed a rigging cable and pulled with all of his might. He looked up to see a silver triangular piece of fabric floating down from the sky before landing at his feet. Mr. Tucker let go of his rope for a moment to retrieve the fabric. There was dried cement along the edges, and Mr. Tucker quickly realized what he was holding. Lifting it up, he turned to the nearest rigger and called, "There is one of the patches dropping off all ready!"[226]

Captain Mabry ordered the engines on before idled. A collective sigh of relief spread among the crew members when the engines started without hesitation. Despite the forty-eight-degree Fahrenheit temperature and mild northerly winds,[227] there wasn't a frozen engine to be found.

Major Thornell came to the doorway of the passenger cabin and gestured to Majors Vautsmeier and Reardon. "Come on down here where you can see the wheels go round!"[228] he eagerly called to the two men. Like two impetuous children, they made a dash for the control cabin.

ROMA leaving the ground on February 21, 1922, with the *A-4* blimp in the background. *Air Combat Command History Office, Joint Base Langley-Eustis.*

Captain Mabry leaned from the window and called, "All right, men, let her go!"

The handling crew chief called back, "Good flying, sir!" Mabry flashed his signature smile and pulled his head back into the window. The men let the ropes slip slowly from their hands, removing *ROMA* and her crew from their last attachment to the Earth below.

ROMA began to rise far more quickly than her crew anticipated. She violently pitched forward as the stern lifted at a much more rapid pace than the bow. "Everyone aft!" Captain Mabry immediately ordered the crew.[229] It was a natural instinct for every lighter-than-air man to balance his ship by any means necessary. Every man who was able moved as far to the back of the ship as possible. Anything that was loose slid forward as the men gripped whatever they could to stay upright.

Sergeant Chapman desperately worked to valve hydrogen gas to the forward compartments, attempting to level the ship out. "The forward air scoop[230] is sticking...damn thing is going up nose heavy, sir!"[231]

"Sergeant, wait! She'll gain her horizontal equilibrium. Just watch," Captain Mabry assured Chapman.[232] Mabry tried to maintain his footing while keeping a calm composure in front of his men. Chapman and Reed

The last known photograph of *ROMA* before the disaster. *Air Combat Command History Office, Joint Base Langley-Eustis.*

worked in conjunction, releasing gas as Captain Reed pitched the tail fins. Suddenly, the air scoop began working, and a collective sigh was felt throughout the control cabin. When the ship rose to five hundred feet above Langley Field, just as Captain Mabry had predicted, *ROMA* leveled out.[233]

"Men, let's hope that's the last incident. Let's take her out and show what these engines are made of! Take us south, Lieutenant Burt."[234] Mabry ordered, pointing his finger forward. The signal was sent through the ship's telegraph system to bring the engines to cruising speed and each was swiftly brought to 1,100 RPM without incident.

Making their way over the mouth of the Back River, Reed commented in a puzzled voice, "She's a bit more sensitive…responding more readily."[235]

Not looking away from his wheel, Lieutenant Burt replied, "We are not getting as much vibration as we had with the Ansaldos."[236]

Major Thornell nodded his head and, with a grin, turned to Major Reardon and exclaimed, "These Liberties are the stuff!"[237]

Walter McNair inspected his instrument with a most bewildered expression. "My instrument says that we're registering seventy-five miles per hour."

Captain Mabry dashed over, noticing that the needle pointed to the far end of the instrument. "The ship has never traveled that fast before…You better go see Mr. Holmes. I hear he tinkered with instruments in the civilian world.[238] He'll be able to tell if your gadget is working correctly." McNair agreed and disappeared from the cabin.

Private Marion Hill appeared in the doorway, a panicked expression crossing his face. "Sir, compartment 1 isn't taking pressure," he reported.

"Never a moment's rest…Mr. Chapman, find McNally and have him check this out," Mabry sighed.

Sergeant Chapman followed Private Hill forward and found Master Sergeant McNally. They quickly explained the situation, and Master Sergeant McNally jumped into action. He climbed up into the gas compartments, and within only three minutes, he reappeared. McNally turned to Chapman and said, "It's taking pressure again but watch it…You will probably have to keep pulling her valve controls."[239]

Despite the stress between the control cabin and the forward compartment, the men on the other parts of the ship were in better spirits. Upon seeing the positive response from his engine, Jethro Beall felt a little more at ease and escaped to the doorway leading to his engine's platform. Majors Reardon and Vautsmeier busied themselves, pestering Captain Reed and Lieutenant Burt with questions about the controls.

When the ship began crossing over Phoebus, Virgil Hoffman left his post and rushed to an open window in the keel. He leaned out and was able to spot his future in-laws' home.[240] He squinted his eyes and was able to make out his beautiful fiancée standing on the ground below. A surge of excitement ran through his body. Virgil's face beamed as he began waving his arms to her. "Stella, I love you!" he screamed. She smiled and waved back, blowing kisses to her beloved. At that moment, all the worry that nestled inside of her body was washed away with pride and enthusiasm for her Virgil.

Ray Hurley, a civilian mechanic from McCook Field, was assigned to train Private Virden Peek on the rear starboard engine. He had never been up in an airship before and felt a great deal of apprehension. In the doorway leading to the outrigger platform, Ray anxiously glanced down at the world passing below him.[241] As the wind tousled his dark hair, he felt fearful that his thin frame wouldn't keep upright and that he would plummet to his death. Ray held a tight grip on the canvas-covered keel to maintain his balance. Private Peek, being quite adept at free standing at such great heights, seemed oblivious to Ray's insecurities. "Listen to the competent clatter of these Liberties! The ship seems to be behaving the best she ever has here in the

States!"[242] he called loudly to Ray, his voice barely penetrating the rattle of the engines and the wind circling around them. Ray motioned for Virden to come back in off the platform. Watching such a young man balance so precariously on the small metal platform only exasperated his sense of vertigo.

Virden obligingly came to the doorway, and the two men sat down together. Peak pointed to the ground, "You see that over there? That's Phoebus. It's near the beach. It's a great place to enjoy the sunshine and meet some pretty girls. And that over there, that's Fortress Monroe. It's pretty old. Apparently it's pretty steeped in history but I couldn't tell ya what the heck it was!"[243] The two men laughed together. Peek's easygoing personality immediately put Ray at ease. Though the two men had never met before Ray came to Langley, they were becoming fast acquaintances.

Lieutenant Burt began circling *ROMA* to pass over the Old Point Comfort lighthouse at Fortress Monroe.[244] The men looked down to see throngs of men, women and children on the ground below waving and cheering.[245] Any sense of worry disintegrated as the men waved back to the adoring crowd below. Little Flores poked his head out of the crow's nest hatch to inspect the top of the envelope. He noticed that the nose of the ship didn't seem quite as taut as it should be. After taking some measurements, he realized that the pressure was off by ten millimeters.[246] While this wouldn't normally seem like a vast disparity, for a ship that was reliant on the pressure within the bag to maintain its very structure, even the slightest change in pressure could affect its every function and the very stability of the ship.

Meanwhile, in the control cabin, Walter Reed felt overcome by his flu symptoms. He began to sweat and clutched his stomach, feeling incredible waves of nausea. He turned to Byron Burt, "Would you relieve me?"[247]

Lieutenant Burt, ever focused, replied, "Just a moment, sir. Allow me to get permission from Captain Mabry." Walter nodded back, unwilling and unable to put up any sort of argument.

Overhearing the exchange, Mabry approached the men and said, "Walt, go lay down. You look like hell. Lieutenant, take over as pilot of elevation. Boys, I'm taking the helm." Reed quietly thanked both men and stumbled toward the passenger cabin. Lieutenant Burt slid into Reed's position as Captain Mabry grasped the large directional wheel in his hands; a sense of pride washed over him. Mabry then gave the order for Burt to dynamically raise the ship's altitude another 150 feet. *ROMA* was only about 500 feet in the air.[248]

ROMA crossed over the water and made her way over Willoughby Spit. The small rock jetty and peninsula were filled with a large crowd of admirers, waving to the ship above them.[249]

In the control cabin, Majors Reardon and Vautsmeier continued to hound their shipmates with questions. "Hey, Burt," Vautsmeier said, "Let me handle the wheel, if only for a minute. You know, so I can say that I had my hands on the controls." Burt replied with only an obligatory nod yet refused to let either man take over. He felt that the ship was behaving far too sensitively to let an inexperienced man take the controls. Burt was also concerned because the ship was just flying lower than usual.

Virden Peek continued to point out different landmarks to Ray Hurley. Having never been to this part of the country, Ray felt it was wonderful seeing it from this perspective. He begun to understand the appeal flight had on men like Private Peek—the feeling of the wind blowing through your hair and against your face, to be free of the binds that tie to the ground and simply gaze down on the world as a bird would. This must be what true freedom felt like. They passed over the Naval Station, with the grey torpedo destroyers bobbing in the water and sailors scurrying about the decks. [250] Ray's apprehension turned to exquisite exuberance.

At that moment, he felt a slight tremor and the ship slightly tilted. Looking to Private Peek for a reaction but seeing none, Ray figured that this was just normal airship behavior.

In the control cabin, Master Sergeant Harry Chapman watched the glass manometers and pressure gauges intensely. He noticed something alarming yet not unexpected. "Sir, compartment 1 is losing pressure again," Master Sergeant Chapman reported to Mabry.

"Go find McNally," Mabry replied.

"Aye, sir," Chapman said as he disappeared from the control cabin.

Mabry began chatting with Major Thornell, and Walter McNair buzzed about the cabin with his instrument, not saying a word to anyone. Burt continued to fend off the nagging of the two majors, attempting to stay focused on his job. It was 2:10 p.m., and *ROMA* was closing in on the Army Quartermaster Depot next to the Elizabeth River. Lieutenant Burt pulled the ropes of the elevator controls, which felt loose in his hands. He pulled again, but still no response. Burt turned his attention to the altitude wheel and gave it a spin. He felt absolutely no resistance; it was dead in his hands. The wheel spun freely with just the tap of his finger.[251] For the first time, he felt panic spread through his body. With wide eyes, he looked up and snapped his head toward Captain Mabry. "She won't respond, Captain!" [252]

CHAPTER 10

THE FALL

"My God, boys!"
—*Captain Dale Mabry*[253]

Alberto Flores pulled his head out of the hatch of the crow's nest to check the top of the bag's pressure once again. He didn't need to take measurements to see the nose of the ship had flattened even more. "I need to tell the Captain!" Little Flores climbed down the ladder leading to the passageway to the main part of the ship. His feet slipped beneath him as he quickly descended the fifteen feet into the bowels of the ship.[254] Upon reaching the passageway, he found it pinched closed so that it was no longer passable.[255] Panic stricken and with nowhere else to go, he climbed back up the ladder.[256] Whatever was going to happen, he had no choice but to stay at the top of the ship.

Meanwhile, Master Sergeant Harry Chapman reached the other end of the same passageway, also finding it not passable.[257] "What the hell?!" he exclaimed. He turned around to find supply officer Lieutenant Ambrose Clinton. Without a word, he gestured for Harry to follow him. The two ran to the rear of the ship, ignoring curious glances and questions from their crewmates along the way.

On the ground, the cheers and adulation turned to utterances of helpless concern. In a nearby neighborhood, seven-year-old Paul Mingee was playing in his backyard while his mother hung the wash to dry. They had timed it perfectly to be able to watch *ROMA* pass overhead. He noticed that the nose of the airship seemed to be crinkled. "Momma, come look!" he exclaimed, pointing to the sky.

His mother dashed over and looked from her son to the sky. Her face fell as she cupped her hand over her mouth. Wrapping her arm around her son, Mrs. Mingee said under her breath, "Those poor men! Oh, those poor men!"[258]

At the Pine Beach Hotel, navy airship pilot James Lawrence stood with his friend A.P. Schneider on the porch. Lawrence turned to Schneider and said, "Their box rudder is coming loose. I hope they can see it." Schneider nodded in agreement.[259]

In the air, Master Sergeant Chapman and Lieutenant Clinton stopped to glance out of a window. They noted that the envelope's fabric seemed

A portrait of Lieutenant Ambrose V. Clinton in uniform. *Hampton History Museum 1987-18-122.*

loose, but of more pressing concern, the box kite rudder was angled in such a way that the ship would descend at a forty-five-degree angle. Upon this discovery, they found that the keel had begun to buckle between the second and third sections. "I have to tell Major Thornell!" Chapman exclaimed and dashed off before Clinton could reply.[260]

Having no idea what Master Sergeant Chapman and Lieutenant Clinton had just discovered, Lieutenant Byron Burt was desperately trying to get the rudder controls to respond in the control cabin. Major John Thornell jumped into action and began pulling the elevation ropes. Captain Mabry gripped desperately onto the directional wheel, the expression on his face confused and panic-stricken. The men in the passenger cabin were alarmed after hearing all of the commotion from the nearby control cabin.[261]

At the Army Quartermaster Depot, employees began trickling outside to watch the ship fly over. Expressions of curiosity quickly turned to horror. An employee at the depot, Graham Dalton, turned to the man closest to him and commented, "It looks like she's going to turn over!"[262]

On board *ROMA*, Walter Reed jumped to his feet and dashed into the control cabin.[263] When he saw the expression on Captain Mabry's face, he exclaimed, "Dale, what do we do next?"

Mabry didn't respond. Instead, Lieutenant Burt immediately jumped into action. "Put out the order to the engineers to cut their motors!" he ordered.[264] Reed nodded and sent the message to the engineers via the telegraph system within the ship.[265]

Sergeant Joe Biedenbach was standing in the doorway to the outrigger catwalk to the starboard center engine, making a notation in his engine's log. He heard the faint sound of his telegraph ringing and glanced over to it. The message read: CUT YOUR MOTORS.[266] He immediately jumped into action. Soon enough, Jethro Beall had idled his engine on the opposite side of the ship, as had Staff Sergeant Louis Hilliard and Private Virden Peak on the rear motors. However, the telegraph system hadn't reached the forward pair of motors, and they were still turning.[267] The ship seemed to idle just for a moment.[268]

Dale Mabry shook his panic and a serious expression crossed his face. "Get her to the Norfolk Country Club!" he ordered, referring to a nearby country club with a grassy field surrounding it.[269] It became obvious to Captain Mabry that he needed to make plans for a forced landing and set the ship down as soon and as safely as possible.

Lieutenant Burt screamed back, "Sir, she still won't respond!"[270]

It was then that the men first felt the ship tilt, nose first. "Sir, we're losing altitude!" Reed yelled.[271]

"Drop whatever you can—lighten the load! We have to even this bloody ship out!" Mabry ordered.[272] Without a central or quick way to drop the ship's ballast, the forty-five men on board began throwing whatever they could get their hands on—batteries, tools, spare parts—out the windows.[273]

The witnesses on the ground seemed frozen in a moment of terror as they saw the ship angle downward. At the Army Quartermaster Depot, Graham Dalton screamed, "There's something wrong with that ship! She's going to crash!"[274]

It became obvious that there was no way that the ship would make it to the empty field around the Norfolk Country Club.[275] This was no longer about saving the ship; it was about saving themselves. Now they could only hope to make it far enough to reach the relative safety of the Elizabeth River before crashing. Virden Peek was too insane to think clearly[276] as he fumbled his way inside the keel. Men began losing their footing and slipping toward the bow of the ship.[277] John Reardon slid through the door of the control cabin and into the passenger cabin.[278] Sergeant Joe Biedenbach dashed inside the keel and climbed to the passenger cabin to see if there was anything he could do to help. He found his crewmates still throwing things overboard and falling into one another, and a few were kicking open the door in order to judge how high they were.[279]

The wind beat tremendously into the cabin, stinging their skin and making it difficult for them to breathe. The ship rushed toward a brick smoke stack at incredible speed. All the men in the passenger cabin huddled together in a corner, bracing themselves for a seemingly inevitable impact. The men watched through the open door to see the ship passing by the smoke stack and heaved a collective sigh of relief. Sergeant Biedenbach declared, "Thank God for that!"[280]

The momentary reprieve was broken when Lieutenant Riley crawled toward the door. A look of terror and madness filled his eyes. In a shrill voice, he screamed, "I'm getting out of here!" The men lurched toward him as best they could, protesting that they weren't high enough up for him to safely jump. Reardon glanced away for a second to look out of a window to try to gauge how high up they were. He heard the men screaming and then all he could see outside was a partially opened parachute.[281] *ROMA* was only two hundred feet above the ground.

Workers at nearby factories and businesses went to their windows and to the streets, motionless in their disbelief of what they were seeing. Pointing to the sky, a worker at the Standard Oil Company's plant at nearby Sewell's Point exclaimed, "A man's falling! A man fell from *ROMA*!"[282] It was as if the entire city held their breath, completely helpless as they watched the man fall through the clouds and toward the ground at a terrific speed. When he disappeared beneath the tree line, no one could decide whether or not they saw the parachute completely open. They could only hope for the best.

At the Army Quartermaster Depot, Captain Whitehurst of the Army Engineering Corps joined the others who swarmed from the ramshackle buildings out onto the railroad tracks. Turning to the post's civilian chief electrician, William Jones, he exclaimed, "It looks like the ship is breaking up[283]…She's going to crash!"[284] Without a moment's hesitation, Mr. Jones ran toward Warehouse 11 to reach a motorcycle he kept there.[285] Often tripping, it seemed as though his feet couldn't move fast enough. Remembering the accidents the year before and knowing that a single spark could light the hydrogen gas ablaze, he knew that his best bet to help was to cut off the power at a substation on the other side of the post.[286]

In *ROMA*'s control cabin, Byron Burt looked out the window as the ground was rapidly approaching.[287] Men all around him were screaming in terror; some were praying while others were weeping.[288] It was a terrible, inevitable drop that they had no choice but to endure. Burt took a deep breath, memorizing every detail he could: the speed being indicated on Walter McNair's now abandoned instrument, the position of the controls, where all of the men were and the purity indications. In all of the pandemonium, he knew that every element was important and necessary; he needed to remember every detail that he could.

With the engines now abandoned, Jethro Beall retreated into the ship. He was attempting to find a parachute or anything that could potentially save his life. Through the madness, he came across Virgil Hoffman, attempting to usher everyone he found to the rear of the ship. "We've got to get out of here, Virg! I told everyone this was going to happen! We've got to get out of here!" Jethro exclaimed.

Looking around for anyone else but seeing no one in sight, Virgil nodded and said, "C'mon," and they disappeared within the ship.

In the crow's nest, Alberto Flores had the most frightening perspective of anyone on *ROMA*. Without any knowledge of what was happening inside the ship and the closest parachute two hundred feet out of his reach,[289] he gripped what he could. The world passed by him at a dreadful pace. In his head, he kept going over every scenario that could occur. Maybe Captain Mabry would get the ship up again…Maybe they were going to crash… Maybe the ship would be able to stay in the air just long enough to reach the water for a softer landing…Maybe he should just focus on his own survival.

In the control cabin, Captain Mabry glanced around at the mayhem surrounding him. He was helpless in the fact that he couldn't save his ship. All he could do was steer and hope that by divine providence the ship would make it to the water. He desperately gripped his wheel, knowing that it was a futile gesture. The forward engines were driving them too fast toward the ground. He gulped, realizing that he, too, was helpless. Only a mere twenty seconds had elapsed since the ship first began losing altitude.[290] Failing to control himself any longer, Captain Mabry took a deep breath before letting out one last booming cry that resonated in every ear on the ship: "My God, boys!"[291]

CHAPTER 11

THE CONFLAGRATION

The heavy envelope fell like a pall over us.
—*Major John T. Reardon*[292]

William Jones climbed onto his motorcycle and was able to start it with the first kick. Manic yet focused, he sped away from the warehouse toward the substation. The ground, soft from that morning's rain, slipped in the tread of his tires, and he couldn't seem to move fast enough. Mud sloshed against his pants and splashed on his face. He wiped his eyes with the sleeve of his shirt. He hadn't remembered the substation seeming so far away. The base was covered in power lines. He knew that if there was any way he could help, it would be to cut the power. People rushed by him toward the railroad tracks—soldiers, civilians—but he didn't have time to notice. He heard shrill screams behind him, echoing through the afternoon sky that sounded like the dreadful cry of a banshee. He had to help; he had to keep going. The damn motorcycle couldn't seem to move fast enough. He needed at least a minute to get the power shut down once he got to the substation.[293] There were too many wires and too many electric currents running through the post.

The seconds felt like hours until he reached the substation. He kicked off the motorcycle as if it were a pair of shoes and dashed to the door. He fumbled with the handle for a moment. It was then that he felt a distinct rumbling through the ground, and the glass in the window shook.[294] The deafening sound of a thunderous crash filled his ears. He gasped as he felt tears fill his eyes. He was too late. There was nothing more he wcould do.[295]

Alberto Flores watched the ground come closer and closer to him as he hunched himself into position. He realized that his only chance of survival was to jump. Without a parachute, he knew that the window of opportunity to do so would be very limited.[296] He remained vigilant and ready. He was alone, and there was nothing he could do for his shipmates inside.

Seconds later, the front tip of the copula smashed into the ground. Flores was maybe fifty feet from the ground at best—this was his moment.[297] Like a cat attacking its prey, Little Flores pounced; his arms flailed for a moment as if he were a baby bird learning to fly. Midair, he began to run. He kept running and wouldn't stop.[298] Running was the difference between life or death, and he was choosing life. He barely felt the mud as he landed on the ground below. There was nothing he could do but run. He felt a tremor through the ground, accompanied by an enveloping crash of sound.[299] He kept running and didn't look back.[300]

Byron Burt braced himself, holding on to the open hay window of the control cabin. Everything was happening so fast. He couldn't think objectively; all he could think about was his wife. Emotions welled inside of him. With a tremendous thud, everything in the ship began to fly forward. Without realizing it, Byron found himself flying through the air and onto the ground. Pieces of wreckage were all around him. He was stunned for a moment, not knowing quite what to do. Then it occurred to him that he was alive. He had to get out. Byron pulled the loose canvas and metal framework from on top of his body and crawled out. He felt a tremendous heat against the back of his neck as he stumbled to his feet and ran as fast as he could. He couldn't decide whether or not he was hurt. After he ran about fifty feet, Byron stopped and turned around. *ROMA* was on fire![301] He had to get his shipmates out.

What was left of the inside of the passenger cabin was obscured by black smoke.[302] Major Reardon was thrown onto his back and watched as fire flooded the cabin as water would a sinking ship. He didn't know what to do—maybe he would wait until the fire burned a hole through the fabric so he could get out. Then he heard Lieutenant Clarence Welch yell, "Anybody got a knife?" John pulled off his glove to fumble for one his pocket. Then, fire flashed through the cabin. Instinctually, he closed his eyes and threw his hands over his face. He went to move but his leg was caught on something. It was the framework of the keel! What was he going to do? He felt something tug his arm upward, the wreckage around his leg loosened enough and he was freed.[303]

First responders attempt to douse the flames from *ROMA*'s ruptured fuel and oil tanks after her crash. *J. Sargeant Memorial Collection, Norfolk Public Library.*

"Get out of here and be damn quick about it!"[304] he heard a voice yell.

Outside, Byron Burt turned to run toward the ship. He saw a man crawling from the wreckage. His skin was red and caked with mud. He couldn't recognize who it was. "Burt!" the man exclaimed. Burt ran toward him and pulled the man by the straps of his parachute pack before dragging him to safety. Once they were at a safe enough distance, Burt pulled the man to his feet. It was then that he noticed fire on the back of the man's uniform. He slapped him on the back to put it out. "You're going to be all right!" Burt exclaimed and then ran back towards the ship.[305]

John Reardon stumbled by him. "Burt, they're in there! We've got to get them out!" Reardon exclaimed. Burt nodded and continued toward the ship. He passed another man so completely covered in mud that only his dazed eyes could be seen. Burt couldn't tell who it was.[306] This man seemed fine enough, though. Burt had to keep going.

He then came across Clarence Welch stumbling through the mud. His uniform had a small fire burning on it, but Welch seemed to be moving just fine. Burt decided that Welch was able to help himself.[307] There were still men in the ship. He knew that his time was limited—he had to get his shipmates out.

A third, powerful explosion erupted, knocking Burt off his feet.[308] A conflagration of insurmountable proportions leaped into the sky; it was an inferno of incredible intensity. Burt felt someone pulling at his arm as he stumbled to his feet. It was Captain Lawrence Woods, the commanding officer of the depot. "Lieutenant, we have to go!" Captain Woods exclaimed.

"I need to go back! I need to save them!" Burt screamed back.

"Lieutenant, you'll be killed!"

"I don't care! I have got to save those men!"

"There's nothing more you can do for them! Now, go! That's an order, Lieutenant!"[309] Captain Woods screamed. Burt clenched his jaw. He looked to the fire and then looked back to Captain Woods. He nodded and reluctantly turned away.

Sirens cut through the air as engine companies and ambulances rushed to the scene. Firefighters keenly trained their hoses on the blaze, but to no avail.[310] Witnesses could make out the distinct forms of dark, human-shaped figures in the fire, writhing in agony. Their arms were batting above their heads in a fruitless attempt to escape the fire engulfing their bodies. One let out a screech that pierced through the fire and smoke, pitifully screaming, "Oh, mercy! Mercy!"[311] Within seconds, the fire consumed the figure, his miserable cry dying along with him. In the following moments, the distinct

Firefighting crews desperately attempt to extinguish the flames that engulfed *ROMA*. *Air Combat Command History Office, Joint Base Langley-Eustis.*

sounds of human groans could be heard before dissipating into the crackling flames of the pyre.

Burt spied a group of men huddled by the roadway and ran to them. He recognized Ray Hurley, who seemed relatively unscathed. With a sigh of relief, he wondered how many others had gotten out. Then Burt followed the direction in which Ray was looking, and his initial relief transformed to horror. A half-opened parachute was lying spread out on the cement. Byron's eyes traced the strings of the parachute to see it attached to a man. He was lying on his back, his crushed head tilted to his left side.[312] Ray turned away from the body and toward Burt. "It's Lieutenant Riley," he mumbled. An ambulance pulled up, and the medics worked to cut the parachute from William Riley's body before loading him onto a stretcher[313] and whisking him away. Perhaps it was more wishful thinking than anything else, but no one could quite agree as to whether Riley had actually survived the fall. It would later be determined that when Lieutenant Riley leaped from *ROMA*, he wasn't at a high enough altitude for his parachute to be able to completely open. Because of the nature of his injuries, the doctors examining his remains concluded that he more than likely died on impact. Lieutenant William Riley left behind a wife and a five-week-old daughter.

Men began loading others onto stretchers and into cars and ambulances. Captain Lawrence Woods stood near the vehicles, pencil and a pad of paper in hand.[314] He was taking down the names of every person loaded before sending the vehicle on its way. Byron saw Walter McNair following behind a stretcher and ran to him. "Who is that?" he asked.

McNair replied, "It's Master Sergeant Chapman. He ran back into the fire, and he saved my life. He also saved three other men."[315]

CHAPTER 12

THE SURVIVORS

I'm ready to go home…if the doctors say so.
—Captain Walter J. Reed

Maria Reed was enjoying the merriment of the game of bridge she was playing with a few of her friends. They made lighthearted conversation, and it was a nice reprieve from that morning's argument with her husband. A knock came at the door, and she politely excused herself to answer it. Upon opening the door, she saw Captain William Kepner on the other side. "To what do I owe the pleasure, Mr. Kepner?" she asked.

"It's *ROMA*, Mrs. Reed—she crashed," he told her, fright covering his pallid face. She looked to her friends and grabbed her coat.

"Take me to where I need to go," she insisted. Kepner drove her to the offices at the airship school and directed her to Captains Lawsen and Hasterbrook,[316] who were preparing to fly across the water to the wreckage.

"Is there any news on Walt?" she asked. Her blue eyes were wide with fear. The two captains looked at each other and then to her.

"We don't know," Captain Lawsen told her.

"Take me over there. I need to find my husband!" she insisted.[317] Captains Lawsen and Hasterbrook didn't put up a fight and guided her to the plane. After takeoff, Maria could see smoke filling the air like a dark cataclysm engulfing Hampton Roads. As they drew closer, she could see enormous flames shooting into the sky. She recoiled in horror.

"Where would they have taken the men?" she asked.

An aerial view of *ROMA*'s crash site. *Air Combat Command History Office, Joint Base Langley-Eustis.*

"Mrs. Reed, we don't know who—" Captain Hasterbrook started.

"I don't care! Now tell me where they took the men!" she exclaimed in frustration. It wasn't like her to break her composure.

"They were taken to the Public Health Hospital at the Quartermaster Depot, less than a mile from where the ship crashed.[318] I'll have someone drive you there once we land."

The plane landed at the Naval Air Station, and a young sailor drove Maria to the hospital. The drive was silent and dreadful. She thumbed her fingers in her lap as she watched the swarm of cars go by them en route to the scene. A blackened cloud of smoke had descended on Norfolk, and it was hard for the sailor to see his way through.

The car stopped in front of a small makeshift building. Maria scrambled from the car, barely remembering to thank the young man for driving her. She barreled through the door and immediately came upon a nurse. "The men from *ROMA*—where are they?" she insisted.

"And you are?" the nurse asked, inquisitively.

"I'm Mrs. Reed. My husband was on that ship!" Maria exclaimed, exasperated.

"Wait right here," the nurse replied. She quickly returned with a physician named Dr. Quick in tow.[319]

"Mrs. Reed, I'm Dr. Quick. I'm in charge of caring for the men from *ROMA*. Please follow me," the doctor said and began walking rapidly deeper into the building. Maria remained steadily at his heels.

They entered a room that was filled with a great stench that could only be compared to burned meat. Doctors and nurses stumbled around from gurney to gurney, loudly calling to one another and fumbling through drawers and containers, searching for supplies. Screams and cries of agony came from the patients. Maria tried to scan the faces of the men, but most she could barely recognize past the burns. Her lips trembled, and she clutched her stomach. It was a most horrible scene. She continued to scan the faces until she looked upon a face she knew all too well. With her fear tempered, she exclaimed, "Walt!" and ran to his side. His head and hands were wrapped in bandages and his uniform in pieces around him. His eyes were solemn and his face drawn until he heard her voice. It was like hearing angels casting mercy on him.

"What happened? Are you okay?" she asked him, tears streaming down her cheeks.

"Lieutenant Welch got me out, and Burt pulled me from the wreckage," he said. Reed looked up at Dr. Quick and then back to his wife. "I'm ready to go home, if the doctors say so."

Dr. Quick looked at Captain Reed seriously. "No, Captain, you need to stay here. We need to continue to assess your wounds." In his great exhaustion, Reed simply nodded in acknowledgement.[320]

At the Army Quartermaster Depot, the fire continued to rage. Captains Lawsen and Hasterbrook arrived at the scene and immediately came upon Lieutenant Burt. His face was distressed in a way that neither of them had ever seen before. Before they had a chance to say anything, Burt turned to Hasterbrook and asked, "How did you get over here?"[321]

"We…we took a plane."

"Sir, would one of you mind taking me back to Langley?" Burt asked. He needed to report everything he saw as soon as possible.

"Sure," Captain Lawsen replied, his voice sounding mechanical due to the shock that was filling him.

Captain Woods approached the men and inquired as to why Burt hadn't yet reported to the hospital. Despite his protests, Captain Woods signaled for a man to drive him to be checked over.

The wreckage was still consumed by a great inferno. The framework of what was once the rudder sat propped up between two telephone poles; a broken electric wire dangled free from one of the poles. It was hard to delineate the

RUINS.OF ROMA FEB. 21, 1922.

A close-up view of the remains of *ROMA*'s box kite rudder resting on the high-voltage lines. *J. Sargeant Memorial Collection, Norfolk Public Library.*

shape of *ROMA*. Her rigid framework was crushed and twisted on top of railroad tracks and pig iron.[322] Sailors from the navy base had arrived to help combat the fire and aid in recovery efforts. Reporters began trickling onto the scene, and photographers were busy taking pictures of the ruined dirigible.

Captain Woods called on the general superintendent of the crane service for the depot, Lloyd Bradgor, asking him to order a crane to clear the tracks. Perhaps more victims could be recovered.[323] Grimly, Captain Woods contacted Mr. Rouse, a funeral director in Newport News, to bring hearses to the depot for after the recovery was completed.[324]

At the Public Health Hospital, the men's wounds were being treated by scraping the burned and dead skin from the wounds with a hard-bristled brush. It was an incredibly painful experience. They gripped tightly to the metal poles of the beds and made a game out of seeing who would bend theirs the most.[325] After each scraping session, picric acid was dabbed on the fresh, raw skin, followed by layers of gauze.[326] Some of the men began talking among themselves, trying to figure out what exactly happened and who else had gotten out.

A crane begins removing *ROMA*'s frame from the railroad tracks at the Army Quartermaster Depot. *J. Sargeant Memorial Collection, Norfolk Public Library.*

Ray Hurley searched the faces and was relieved to find Private Virden Peek. They sat down together, and Ray asked, "What happened in there?"

"I looked out and saw that the ship was right close to the ground. Sergeant Hilliard was standing by the instrument board at the rear motor group, and the ship got to such an angle that Sergeant Hilliard fell down about the time the ship crashed. His head caught in the keel, and I tried to get him out but I couldn't. I pulled him with my left hand and held myself up with my right hand. I pulled him so hard that I pulled all the buttons off his shirt. I think he died while I was trying to get him out of the ship. I kept pulling on Sergeant Hilliard trying to get him out..."[327] Virden started to weep. Ray put his hand on his friend's shoulder.

"You did what you could," he said.

"I had him by the collar of his shirt. The keel had Sergeant Hilliard right at the throat...I think he died while I was trying to get him out."

On the other side of the room, Lieutenant Clarence Welch sat near Walter Reed to keep him company. "I heard that Chapman might not make it...Joe Biedenbach pulled him from the ship after Chapman saved a bunch of men. I also heard that Charles Dworack is in a bad way," Lieutenant Welch said.[328]

"Mr. Dworack made it out, too?" Reed asked.

Welch nodded. He looked around and counted, "There are seven of us here. We know that Little Flores and Burt made it out. Chapman and Dworack must be somewhere else."

"So, that's eleven. I heard that Riley didn't make it. I wonder where everyone else is?" Reed asked.[329] Welch could only shake his head.

A fedora-capped man stealthily walked into the room, pencil and paper in hand. He began moving from man to man, asking questions about the experience. It quickly became common knowledge that this man was a reporter. In all of the confusion, he must have slipped past the nurses. While speaking to John Reardon, Reed called over to him. "Hey, what became of the other boys? Did they get out?"

The reporter stopped, his eyes growing wide, and the expression on his face dropped. He looked at Reed for a moment before replying. "No…no one else got out. You are the only survivors."

The room became silent as the terrible truth fell upon them. Ray Hurley shook his head, and all he could utter was, "Awful!"[330]

Only eleven men out of forty-five had survived.

Only eleven.

CHAPTER 13

THIRTY-FOUR LOST SOULS

I was in for a greater test of my composure when I visited the funeral home…I have never forgotten that night.
—Lieutenant General William Kepner (USAF, ret.)[331]

Lieutenant Burt was soon cleared to return to the Army Quartermaster Depot. Except for very minor burns, he managed to make it out of the crash physically unscathed. He arrived to find more people having swarmed onto the scene. Reporters were trying to speak to him, but he walked past them as if he were oblivious to their attention.[332] A squadron of square black hearses was pulled along the side streets near the wreckage. Mr. Rouse, the funeral director from Newport News, called on several of his colleagues to bring hearses from their mortuaries to aid in the effort.[333] A large crane was being inched closer and closer to what was left of *ROMA*. Rescue parties were being assembled nearby. Captain George B. West of the Medical Corps stationed at the depot was put in charge of what was first touted as a rescue attempt but quickly became a recovery effort.[334]

The fire, though still burning, seemed to have tempered a great bit to a dull spark and a cloud of gray-white smoke. For the combined efforts of the fire companies, this had been no small feat. Large puddles of water lay on the ground, and gasoline moved in large rainbow cascades along the surface. Byron stood as close to his ship as he possibly could, near a pile of pig iron that was stacked near the wreckage. Various girders and pieces of the structure lay mangled in front of him. He knew that inside that mess of twisted metal

Recovery workers search for human remains as the wreckage of *ROMA* is removed from the Quartermaster Depot. *J. Sargeant Memorial Collection, Norfolk Public Library.*

laid his shipmates—his friends. He hunched inside his jacket, placing his ash-covered hands in his pockets. Shifting in his boots on the rubble, he couldn't shake the sadness and disbelief that consumed him. Overcome, he sought out Captain Lawsen so he could return to Langley.

The crane moved its mighty arm toward the wreckage, its hook lifting the twisted metal from the ground. Shreds of scorched silvery silken canvas flapped in the breeze. Captain West began directing men to help. He tasked the depot's chief timekeeper, Herman Dwyer, with the dark task of helping find means to identify each body as it was recovered from the ruins.[335]

The funeral directors brought white sheets to the men as they climbed through the debris. The scene was chaotic, making it difficult to keep curious onlookers away. It was a gruesome and wretched task ahead of these men. The first victim was found: his face and skin burned beyond recognition, most of his uniform charred away. All that was recognizable was a gold watch. With a deep gulp, Mr. Dwyer tied the watch to the man,[336] whoever he may have been.

The sun was beginning to set in the sky, the various hues of oranges and reds replacing the flames that had earlier consumed it. The smoke continued

Lieutenant Byron T. Burt stoically looks at the smoldering remains of *ROMA* after the crash. *Air Combat Command History Office, Joint Base Langley-Eustis.*

Workers from the Quartermaster Depot and surrounding areas extinguish the last flames from the fire that consumed *ROMA*. *J. Sargeant Memorial Collection, Norfolk Public Library.*

Remains are carefully removed from *ROMA*'s twisted wreckage. *Air Combat Command History Office, Joint Base Langley-Eustis.*

to billow into a mass with the clouds above. Some of the men recovered were obviously deceased—missing limbs, heads shrunken by the extreme heat—while others raised a momentary hope for life until it was plain to see that the man the body once belonged to was gone.[337] Each one was wrapped in a clean white sheet and then unceremoniously loaded into a hearse. One by one, Mr. Dwyer tied trinkets to each man: rings, rank insignias, coins, a silver cigarette case. Captain West made an initial assessment as to cause of death. He surmised that some had died in the crash and others asphyxiated while some burned to their deaths. However, it was impossible to truly make a definitive answer one way or another.[338]

It wasn't until close to 6:30 p.m. that the last body was pulled from the wreckage. The man's uniform was badly singed, and his face was burned away. However, his hands were locked in a fiery grip to the navigation wheel of *ROMA*.[339] He was gently wrapped in a sheet like the others and placed in a hearse.

The loud clatter of the hearses echoed in the still February night as they drove in a funeral procession through the streets of Norfolk. When they reached a dock, they were met by a ferry that Captain Woods had chartered.[340] The thirty-four men who had perished that fateful day made their way back across the water that they had flown over just a few hours earlier.

Once word had reached Langley, loved ones of the crew began to gather. Stella Hoover waited patiently, praying and hoping against all hope that her Virgil would safely come home to her. She thought of how his smiling face had looked down on her just a few hours earlier and how impossible it seemed that he might be gone. He couldn't be gone. *He's safe*, she thought. She just had to keep telling herself that he was all right to keep her heart from disintegrating.

Captain William Kepner and four other men from Langley were ordered to the Rouse Funeral Home to help positively identify the victims.[341] He couldn't believe the painful reality. *ROMA* was gone—a ship he had been slated to fly on just hours earlier. It was a grim twist of fate if there ever was one. The men arrived at the funeral home around 8:00 p.m. A crowd was gathered on the sidewalk across the street, craning their necks to catch a glimpse of whatever they could. Meeting the men was a military dentist and a physician from Newport News, Dr. Jesse H. Mabry. He was there to find his brother.[342] Mr. Rouse briefed the men on what they were about to see, but no amount of training or experience could prepare them for the task ahead.

Upon walking into the room, the stench of burnt flesh overcame them. A young sergeant and a lieutenant ran from the room and out into the street, clutching their bodies and gasping for breath.[343] The onlookers on the sidewalk watched the men with great curiosity and began asking them questions. Neither man could speak as they sought respite on the concrete.

Inside the funeral home, the task was one of horror. The details of most of the faces were burned away except for general outlines. Hands, feet and fingers were missing; skin was scorched to a charred crisp.[344] The positions of each seemed to be recoiling into some sort of silent, eternal terror of the hell they must have endured in the last moments of their lives.

The personal effects that Mr. Dwyer had tied to each victim earlier proved beneficial in the arduous identification process. William Kepner knew each of these men so well. A few times, he choked back tears and stepped from the room to regain his composure. Mr. Rouse joined him on one occasion. "All my friends and my recent boon companions,"[345] Kepner whispered. Once the victims were identified, they received labels with their names. The contents from their pockets and other personal effects were packed away in boxes to be given to their loved ones.

The group worked diligently until the final body was unwrapped at close to 1:00 a.m. There was the man, still gripping the navigation wheel. Dr. Jesse Mabry pursed his lips together and nodded his head. It was that of his little brother, Captain Dale Mabry. A hero to the end, Captain Mabry never left his post.[346] Dr. Mabry removed his brother's rank insignia from the shreds of what was left of his uniform collar and then searched what was left of his uniform pockets. There he found a scorched twenty-five-cent coin and a twisted metal keychain.[347] Dr. Mabry turned to Mr. Rouse, thanked him and left the room. This would be the first of many long nights ahead for both those who survived and the loved ones who were left behind.

CHAPTER 14

THE DEATH ROLL

*Their sacrifice was made for the country, for our common welfare, and they
are deserving of all honor for their devotion, which involved their giving
themselves in full measure.*
—*Major General Mason Patrick*[348]

A hush fell like a dark cloak over Hampton Roads on the night of February
21, 1922. It was all too unfathomable. That morning, forty-five men
ascended to the sky. By nightfall, there were more than thirty widows and
orphans left behind.[349] When the fire died down to nothingness and the souls of
the dead departed, all that was left were the ruins of what was once a symbol
of magnificence, grace and glory—now virtually unrecognizable in the clutter
of twisted metal and cloth. The broken, battered funeral pyre that had formerly
been *ROMA* was a lonely sight as she lay barren in the cold February night on
the railroad tracks of the Army Quartermaster Depot.

For the survivors, their doctors made it their goal to stabilize them as
quickly as possible to transfer them to the army hospital at Fortress Monroe.
However, doctors were uncertain as to whether Master Sergeant Harry
Chapman and Mr. Charles Dworack would survive.[350]

The U.S. Army Air Service had already begun its investigation, assigning
Majors Davenport Johnson, John H. Jouett Jr. and Joseph McNarney from
Langley Field with the delicate task.[351] They flew to the scene, examining the
wreck for themselves and taking note of the witnesses with whom they would
need to speak.

ROMA's ruins after her crash at the Army Quartermaster Depot in Norfolk, Virginia. *National Museum of the U.S. Air Force.*

At Langley Field, Lieutenant Colonel Albert Fisher stayed in his office all night. As word came from Rouse Funeral Home of the positive identification of each man, he painstakingly drafted a telegram to the families of the victims.[352] The faces and memories of the dead passed through Lieutenant Colonel Fisher's mind. He had known all of these men. They were the best and brightest of the U.S. Army Air Service's lighter-than-air program, and now they were gone in the blink of an eye. They were more than airmen; they were sons, brothers, husbands, fathers, fiancés, friends and companions.

The telegrams to the families of the junior noncommissioned officers were rather curt and cold, stating: "We regret to advise you that your son met his death as a result of an accident to the dirigible *ROMA*. Please advise whether you desire us to ship his body to you at the Government's expense."[353]

However, to the bereaved families of the senior enlisted men and the officers, the telegram was far more formal and filled with the resounding grief that blanketed the air service. Fisher signed each telegram on behalf of Major General Mason Patrick.[354]

In the very early hours of February 22, 1922, Marie Thornell was getting settled into their new home in Washington, D.C. John was scheduled to be

123

home later that day, so she busied herself preparing for his return. Even at this late hour, she was awake with anticipation. She heard a knock on the door and found a messenger on the front stoop. He handed her an envelope and silently walked away. Having not had time to read the paper or answer the phone that day, she felt a sense of dread in her stomach. Marie closed the door and opened the telegram. It read:

Dear Mrs. Thornell:

When the airship ROMA crashed and was destroyed this sad accident cost the lives of the very best men in the Air Service. Your husband was among those who were killed.

In the name of the entire Air Service, I send you our heartfelt sympathy. All of these men were working to develop air craft and its use in order that this country of ours might be better defended in time of need. They were pioneers in the effort to navigate the air and in the course of their work they laid down their lives. Their sacrifice was made for the country, for our common welfare, and they are deserving of all honor for their devotion, which involved their giving themselves in full measure. We shall ever keep alive the memory of those brave men, and again assure you that we sympathize with you most sincerely in your own sorrow for the loss of him who was dear to you.

With deepest regards,
Major General Mason Patrick
Chief, U.S. Army Air Service[355]

At Langley Field, Stella Hoover refused to leave until she knew where her Virgil was. None of the officers would answer her questions or even speak to her because she wasn't considered family. At close to 2:00 a.m., Captain William Kepner walked into the airship school, carrying a box containing some of the personal belongings of the dead. He saw Stella sitting near the entrance, her large eyes filled with fright and worry. Though officers and enlisted men didn't fraternize, Virgil had been one of the better liked and most trusted of the noncommissioned officers. Kepner had always enjoyed conversing with Virgil and recognized Stella from a few of those occasions. He knelt down in front of her and asked, "You're Sergeant Hoffman's girl, right?"

She nodded and asked meekly, "Where is Virg…Sergeant Hoffman?" Kepner gazed down at his feet as he couldn't look this young woman in the eyes and tell her the awful truth.

After a grueling moment of silence, Kepner turned to the box and pulled out a smaller one, labeled: "SGT. HOFFMAN, V." and placed it in her shaking hands. "I'm so sorry, Miss. I think he would want you to have these," Kepner mumbled before walking away. Stella slowly opened the box, and inside she found Virgil's matchbox and his silver cigarette case, tarnished and stained by smoke. She opened up the case, finding five neatly rolled, untouched cigarettes sitting upright. Her breathless cry cut like a dagger through the silence of the night.[356]

Around the United States, telegrams began arriving at the doors of the family members of the victims. The early edition of the nation's papers arrived on doorsteps and sold on street corners, each front page's headlines detailing the devastating loss. Once again, readers flocked to newsstands to read of *ROMA* but this time in utter disbelief. How could a ship so grand fall from grace, taking with her so many lives? *ROMA* and her crew epitomized the reinvention of American aviation after the war— but now they were lost forever.

News quickly spread around the world, and notes of condolences from leaders of other nations began being printed in the newspapers, including the sympathies of the German government.[357] Smiling, uniform-clad images of the victims filled the headlines beside photos of the wreckage. Personal stories and anecdotes from anyone who may have had even the most minor encounter with the victims were printed. While the nation was bound together in collective grief, the press salivated over every small detail it could put in newsprint about the victims, the ship and the crash.

THE NAMES OF THOSE LOST WITH *ROMA*

Major John G. Thornell
Major Walter W. Vautsmeier
Captain Dale Mabry
Captain George D. Watts
Captain Allan McFarland
Captain Frederick Durrschmidt
Lieutenant John R. Hall

Lieutenant Wallace C. Burns
Lieutenant William E. Riley
Lieutenant Clifford E. Smythe
Lieutenant Wallace C. Cummings
Lieutenant Ambrose V. Clinton
Lieutenant Harold K. Hine
Master Sergeant Roger McNally
Master Sergeant James Murray
Master Sergeant Homer Gorby
Technical Sergeant Lee M. Harris
Staff Sergeant Virgil Hoffman
Staff Sergeant Marion "Jethro" Beall
Staff Sergeant Louis Hilliard
Staff Sergeant Edward M. Schumacher
Staff Sergeant James M. Holmes
Sergeant Thomas Yarbrough
Sergeant William J. Ryan
Corporal Irbey Hevron
Private First Class Gus Kingston
Private First Class Marion Hill
Private Theron N. Blakely
Private John E. Thompson
Walter W. Stryker
Robert J. Hanson
William O'Loughlin
Thomas Harriman
Charles W. Schulenburg

CHAPTER 15

THE INVESTIGATION

We are not merely going to try to find out what happened; we will find out.
—Major General Mason Patrick[358]

The break of dawn on February 22 was met with tears, horror and disbelief. Headlines across the world continued to read of *ROMA*'s demise and the martyrdom of her brave crew. Snapshots of rescuers lifting white sheet–covered bodies from the wreckage were printed alongside the smiling faces of the forlorn lost. Graphic depictions and witness accounts freckled controversial witness statements and questions of *ROMA*'s faults. Stories about *ZR-2*'s crash were resurrected, and accusations ran wild over the use of hydrogen and the government's decision not to procure helium for *ROMA*.

The City of Newport News hurried to plan an elaborate public memorial service befitting of the sacrifice of the fallen men. At Rouse Funeral Home, news from the families trickled in about how they wanted to receive their men. Some hurried to Hampton Roads to claim their lost heroes while others waited at home for the pending arrival of a flag-draped casket bearing the charred remains of their loved one. Each man was to be accompanied home by another soldier, usually a friend or personal acquaintance. True to his word, Corporal Nathan Curro volunteered to escort his friend Corporal Irbey Hevron back home to Indiana.[359]

When family members arrived at the funeral home, Captain William Kepner remained on hand to help the bereaved.

"I need to see my boy," one father insisted.

Rescue workers search for victims in the collapsed keel behind one of *ROMA*'s Liberty motors. *J. Sargeant Memorial Collection, Norfolk Public Library.*

Kepner placed his hand on the man's shoulder and the other on the lid of the casket. "Sir, I don't think that would be a good idea. I assure you that this is your son."

"Are you certain? He has a scar on his hip. He got it in an accident when he was a child. Did you see the scar?"

The father's eyes penetrated deep into Kepner's soul. There was a look of panic and disbelief on the man's face. "Sir, there is no pelvis…" Kepner said, before trailing off.[360] The man collapsed next to the coffin of his son; the look on his face transformed from disbelief to a horrific, hollowing anguish. No parent should ever have to bury their child.

Answers were being demanded by and of the government and the U.S. Army Air Service. The Italian government promptly sent Lieutenant Colonel Alessandro Guidoni to conduct a separate investigation to absolve themselves of blame. Engineers and soldiers from Langley Field were detailed to the scene to begin the grizzly task of climbing through the wreckage to remove any human remains that were still twisted in the wreckage. There they found charred arms, legs and other extremities that had been gruesomely torn from the bodies and could not be individually identified.[361]

Souvenir poachers gained access to the ruins, tearing bits of the silvery envelope and stuffing them into their pockets.[362] A local boy, Louis Hitchings, rode his bicycle to the wreckage and found one of the ship's wheels mixed in the rubble. When he tried to leave with it, he was caught by a soldier guarding the wreckage and was forced to leave the trophy behind.[363]

Morbid curiosity consumed locals, including three hundred state legislators who were visiting Norfolk. That morning, they left the downtown area of the city in a motorcade of 130 cars, led by Governor and Mrs. Trinkle, as well as the city's mayor, Albert Roper, and his wife. While they were en route to the navy base to watch a ceremonial dress parade, the motorcade made a detour into the depot to see the wreckage as an unplanned "inspection."[364] Despite the circumstances of the day before, the ceremonial dress parade continued on as planned.

Nine of the survivors were soon stable enough to ferry across the river to the hospital at Fortress Monroe. However, Master Sergeant Harry Chapman and Charles Dworack would be left behind. Harry's burns were incredibly severe, and Dworack still wasn't expected to survive. The doctors and nurses discovered that his breathing suffered because he had swallowed flames and inhaled gas fumes while in the confines of *ROMA*'s great conflagration.[365] Their fellow survivors waited with bated breath for news of the two men, though maintaining a façade of cheerful gratitude for their own survival. The families of the survivors also began to arrive in Hampton Roads, including Captain Walter Reed's father from Scarsdale, New York.[366]

Major General Mason Patrick arrived at Langley Field to aid in the investigation while also under pressure to furnish answers to both the government and public. No time was to be wasted in this investigation. Majors Johnson, Jouett and McNarney went to the hospital to press the survivors for their own accounts of what had happened the day before. The primary questions were what caused the nose to collapse, what made the rudder drop, if the ship was on fire prior to crashing and what caused the fire.

The first man to be interviewed was Lieutenant Clarence Welch. He was duly sworn in and advised of his protection against self-incrimination under the provisions of the Twenty-Fourth Article of War. The air felt stale and dense in the quiet hush. Specks of dust were seen floating about the room, and sun streamed through the window. Clarence's head throbbed as he found little comfort in the tightly wound bandages around it. He sat with his arms crossed; his hands were coiled in mittens of gauze. He hugged his dressing gown close to him to cover the hospital clothes that shielded his battered body. His expression was distant, and his eyes conveyed the sadness that occupied his soul. He barely heard someone say to him, "State your name, rank, organization and station."[367]

"C.H. Welch, first lieutenant, Air Service, Langley Field, Virginia," he mumbled in reply.

"Do you fully understand your rights under the Twenty-fourth Article of War?" Major Johnson asked.

Welch nodded. "Yes, sir."

"What were your duties on board the airship *ROMA*?" asked Major McNarney.

Welch continued to gaze into the distance. "I was a student passenger," he replied. General questions followed about the ship's altitude and general observations. Clarence's stare continued to remain fixed on an unknown point of the room.

"When did you first notice the fire?" Major Jouett asked. For the first time, Welch's eyes darted to the three majors and sprang to life with a vibrant fury.

"After the crash," he said.

"How long after the crash?"

"It must have been at least fifteen seconds. Just prior to the fire, there were three distinctive explosions that I remember very clearly. I could hear the fire, and I called out for someone to give me a knife so that I could cut my way out of the ship." Gesturing over to Captain Reed, he said, "He gave me the knife and it then came to me that many of the passengers and crew were entangled in the wreckage. There was nothing for me to do but lie there and wait for the fire to burn the fabric. I questioned whether the fire would reach me from above or below first. I pushed a hole in it and got out," Welch said, the ferocity of his voice resonating in the room. The majors stared at him for a moment, attempting to regain their own neutral expressions.

Major Johnson asked him, "In your opinion, if it had not been for the fire, the greater majority of the passengers would have been saved?"

"If there had been no fire, I think almost everyone would have been saved. Some may have been injured from the jolt in the forward part of the ship, but nevertheless they would have been saved," he replied.

"When we say 'fire,' it is necessary to differentiate between the gasoline fire and the hydrogen fire. When you say 'fire,' which one do you mean?" asked Major Jouett.

"I refer to the hydrogen fire. The gasoline fire would undoubtedly have consumed the ship, but there would have been more time to have saved everyone," Welch sternly replied, gritting his teeth together.

There was a pause and then Major McNarney asked, "If there had been a non-burning gas in the envelope, do you think the ship would have burned any slower?"

Welch stared daggers into the majors and took a deep breath. The faces of his fallen shipmates raged through his mind, and even beneath the bandages on his hands, he felt his fists start to ball together. "I am *sure* the ship would not have burned so quickly but what a great number of men could have been saved and an attempt made to save the others. Captain Reed, Lieutenant Burt and the others barely had time to get out themselves and had *no* possible chance to help anyone else!" he exclaimed.

There was another pause. "Did you see Lieutenant Riley just prior to the accident?" asked Major Jouett.

Welch sighed and looked down at his bandaged hands. "Yes, sir," he mumbled.

"Did he have on a parachute?" asked Major Johnson.

Welch nodded and then once again gazed off into an unknown space. "I am sure he had on a parachute. I helped him put it on before we left the ground. The report that the ship was on fire in the air is all folly for I am sure there was no fire in the ship until after the crash. I am quite sure of this or I would not have made the statement." Clarence Welch choked back tears in an effort to regain his composure.

Walter Reed was considered a key witness in the proceedings, since he had not only been with *ROMA* since Italy and was her pilot but had also overseen the reconstruction of the ship after her arrival at Langley.

"I believe the nose section of the ship lost pressure and gave in. This had a tendency to push the keel section back. The cables, which stay the box kite, which are attached to the bag, in turn gave and we dived. I don't believe the dive would have occurred had not the stay ropes and control cable become loosened when the nose gave way. The keel section of the ship is constructed so that if the nose should give away, the entire keel would be injured to such an extent that it would cause the box-kite to become loosened and drop to an angle which would give the ship a tendency a dive. Of course, this is my own theory and I do not know it from fact," he told the board.

"How much experience have you had with airships?" asked Major Johnson.

"Continuously since March 1919," he replied.

Sergeant Joe Biedenbach was then sworn in. "What is your impression of the accident?" Major Jouett asked him.

"The ship nosed down, and we hit the ground at an angle of about forty-five degrees," Joe replied with an unembellished tone.

"What were you doing at the time?" asked Major McNarney.

"I was inside the keel at the time the telegraph rang. I idled my motor. About that time, I was thrown from the ship on the right-hand side. I struck the ground almost fifty feet from the ship. Chapman was about halfway between me and the ship and on fire. I went back and pulled him out," Joe replied, the timbre of his voice growing increasingly quiet.

After speaking with the men, the majors quietly left the hospital and made their way the short distance to what was left of the wreckage. It was scattered and torn in pieces along the railroad tracks. The shattered, blackened pieces of *ROMA* were virtually unrecognizable. Partially destroyed parachutes were intertwined with pieces of the ruins, distinctively cut where they had been torn from the bodies of the men who had worn them just the day before. A great many men were present, quietly speaking among themselves, while others continued to climb through what remained of the great ship, attempting to find whatever was left behind. The board began moving from person to person, asking after their own observations from the day before. Witnesses spoke of how the ship's copula appeared collapsed and loose in the wind, the rudder hanging from the ship at a forty-five-degree angle, the terrific speed with which she barreled toward the ground, the horror of watching Lieutenant Riley fall from the ship to the ground and the tremendous, incredible fire.

In an interview with Willie J. Morris, the chief of the fire department at the base, Major Johnson asked him, "Did you see any fire in the ship before it struck the ground?"

"I am positive that there was no fire in that ship until she hit those wires," he said, pointing to the broken telegraph wire, dangling above the wreckage. "When the ship started down, I did not think for a minute that she was going to crash. I was looking at the ship from the stern, and I know there was no fire until the crash."

"You are sure you would have seen a fire on the ship prior to the crash if there had been one?" asked Major Jouett.

"I am positive. I have been on duty with the fire department for the past eighteen years, and I have been chief of this department for five years. I know there was no fire before the ship crashed," Mr. Morris emphasized.

A young engineer and draftsman, James Gallery, was interviewed. With pencil and paper in hand, he drew for the majors his interpretation of the angle of the rudder. "Was there any fire on the ship before the crash?" asked Major Johnson.

"I am very positive that there was no fire on the ship before the crash. I was looking at the ship very carefully. The fire department was putting water on the fire, which did very little good on account of the gasoline floating on the water. They needed a considerable amount of chemicals. The heat was so intense that the fire would return. I helped a man who I afterwards found out was Lieutenant Welch onto a stretcher from the Public Service Hospital…" Mr. Gallery glanced to his side for a second before continuing. "I rendered all aid I possibly could and everyone at the scene was doing their best. I feel sure that everything humanly possible was done to put out the fire and thereby save the lives of many men."

At 6:30 p.m., the board members adjourned and journeyed back to Langley Field, where they were scheduled to apprise President Warren Harding of their findings.[368]

Lieutenant Colonel Arthur Fisher announced to the media, "The cause of the crash of the *ROMA* was due to failure of the controls to function. The trouble developed at an altitude of five hundred feet and pilots of the ship were unable to prevent its descent. The bag caught fire, following an explosion caused by contact with a 2,500-volt transit line. The large number of casualties was due to the explosion and resultant fire. Otherwise, a much larger number would have escaped serious injury."[369] It was clear: *ROMA* would have crashed no matter what the circumstances. When gas compartment 1 lost pressure due to the deterioration in the envelope, the keel buckled. This severed the elevator controls and caused the rudder to collapse.

Now the attention would be turned to the catalyst that caused the deaths of thirty-three of the men on board: hydrogen.

CHAPTER 16

THE MEMORIAL

They have dropped a stitch in the loom of life, but sometime, somewhere they will catch it again in the life beyond...
—Colonel John R. Saunders, attorney general of Virginia[370]

The newspapers across the country on Thursday, February 23, told of the heroic acts of bravery of *ROMA*'s crew. Sprinkled between the stories were obituaries of the local men who had perished on board. Eyewitness statements and speculations continued to fill the headlines. Letters written by victims about their wariness toward *ROMA* and hydrogen gas were printed, and Jethro Beall's foreshadowing words rang like funeral bells in the black-and-white newsprint.[371] The shock that filled the hearts of every American quickly turned into grief and anger. America needed answers, and they wanted someone to blame for the loss of so many souls. But before the debate could be brought before Congress, the world needed to say goodbye to the thirty-four men who died.

By February 24, most of the men had already departed Rouse's Funeral Home on their final journeys home, led by Captain Allan McFarland.[372] An elaborate memorial service was planned by the City of Newport News. Shops and offices closed for the day, and the population remained home from both work and school. This was a day for the nation to join in its collective grief and bid a final farewell to the thirty-four men who had lost their lives just a few days earlier. Two caskets were taken from Rouse's Funeral Home to St. Paul's Episcopal Church on 34th Street. The light gray stone church sat on a busy intersection. It had a large sanctuary and beautiful blue stained-glass windows. The flag-draped closed coffins of Captain Dale Mabry and Private John E. Thompson spent the morning with mourners filing past them.[373] While Captain Mabry was a natural choice to represent the officers, Private Thompson's body was chosen to represent the crewmen because his was the only one that remained unclaimed.[374] Local civic organizations gathered outside the church to participate in the events of the afternoon. Soldiers and airmen from Langley Field and Fortress Monroe arrived along with the Newport News National Guard Unit, the Huntington Rifles.[375] The widows of Master Sergeant Roger McNally and Technical Sergeant Lee Harris were in attendance, maintaining a composure that was nothing short of admirable. However, none of the eleven survivors were present as they had been issued strict orders to remain silent pending the completion of the formal investigation and the final report submitted to the War Department.

At 3:00 p.m., the sorrowful ringing of the bells of St. Paul's echoed through the streets of Newport News.[376] Along 34th Street, a procession of cars waited behind a flower-encased hearse. Soldiers waited at attention, and the bereaved stood close to one another along the sidewalks. Six white-

A hearse leaving St. Paul's Episcopal Church before proceeding to the Casino Grounds for the public memorial service. *Hampton History Museum 1987-18-92.*

Members of the military and civilians line the streets to watch the funeral procession for those lost on *ROMA*. *Air Combat Command History Office, Joint Base Langley-Eustis (above). Hampton History Museum 1987-18-97 (below).*

gloved soldiers from Langley Field acted as pallbearers and carried the casket of Captain Dale Mabry carefully through the narthex and down the marble steps in front of the church.[377]

The Huntington Rifles presented arms while the pallbearers placed Captain Mabry's casket gently into the hearse.[378] Mrs. McNally and Mrs. Harris were each escorted to a car that was to follow immediately behind.

The funeral procession was led by city and state officials as well as the Newport News Police Department. Following behind was the Fourth Coast Artillery Band from Fortress Monroe. Keeping a somber tempo and pace, they played "Death March" from Handel's *Saul*.[379] The pallbearers walked stoically alongside the hearse.

Two thousand mourners waited along the streets to pay their respects to the fallen. Clad in heavy dark coats that cold afternoon, they stood together in silence. Men bowed their heads and women wiped tears from their eyes, as if the casket contained the earthly remains of their own lost loved ones. The procession slowly made its way to the Casino Grounds at Washington Square.[380] The water lapped along the shore of the James River and bells from nearby ships clanged loudly as they lowered their flags to half-mast. The band played the woeful melody of the hymn "Nearer My God to Thee" as the hearse edged close to a gazebo in the middle of the grassy park along the river's edge.[381]

The hearse came to a quiet halt, and the pallbearers removed Captain Mabry's casket. Without struggle, they placed it gingerly on an awaiting metal catafalque. Two soldiers escorted Mrs. McNally and Mrs. Harris to the gazebo, where two chairs were awaiting them. Mr. Rouse joined the pallbearers who stood at attention near the catafalque as government officials made their way down to speak. The music of the band faded into the silence of the still afternoon air.

Virginia's attorney general, Colonel John R. Saunders, was at the service to represent the government of the Commonwealth of Virginia.[382] He walked in front of the crowd, holding a paper in his hand. Glancing over at the widows and then back to the crowd, he began to deliver a eulogy that he silently hoped would be adequate enough for such an occasion. In a booming voice, he said, "They were men of service and sacrifice…A monument will be erected on the spot where they fell, inscribed with the names of those who died. They have dropped a stitch in the loom of life but sometime, somewhere, they will catch it again in the life beyond. Before you is the body of one who fell. As we leave him, peace to his ashes, peace to his soul."[383]

A small stray dog wandered up next to the catafalque and laid his head down on the ground, as if he, too, understood the sorrowful nature of the

Colonel John Saunders delivering a eulogy at the public memorial service. *Hampton History Museum 1987-18-103.*

circumstance. No one bothered to shoo him away as Chaplain Samuel Smith of Fortress Monroe stood at the head of the casket. Reading from his Bible, he said, "Behold I show you a mystery; we shall not all sleep but we shall be changed. For the trumpet shall sound and the dead shall be raised incorruptible. Then, shall be brought to pass the saying that is written, Death is swallowed up in victory. 'O Death where is thy sting? O Grave where is thy victory?'"[384]

The troops were called to attention. Seven soldiers stood in a line and fired three volleys into the air. Mrs. Harris and Mrs. McNally clutched each other's hands and shook at each crack of the rifle. The faint sound of airplane engines could be heard in the distance as Howard Van Aradale of the Huntington Rifles lifted his bugle to play "Taps." The resolve of the brave widows broke and they wept as planes flew over, manned by friends of the fallen men. Roses fell from their cockpits onto the coffin below.[385]

The crowd dispersed, and Captain Mabry's coffin was removed from the catafalque and then taken back to Rouse's Funeral Home. That night, the remains of Captain Mabry, along with those of Major Walter Vautsmeier, Master Sergeant James Murray and Technical Sergeant Lee Harris, were

A view of the public memorial service for those lost on *ROMA*. *Hampton History Museum 1987-18-101.*

transported to Arlington National Cemetery.[386] Lieutenant Burt readily volunteered to accompany Captain Mabry on his final journey.[387]

In Connecticut, Captain Frederick Durrschmidt's parents received a letter in the mail from an unsigned friend of their son's. It read:

> *I am taking the liberty of writing this letter to the family of my best friend. Although I have not yet met you all, I feel as though at this time we are very close together. The terrible disaster that robbed you of your son took from me the one whom I most loved, from the army an officer and a gentleman, and from the world a man. We slept together, ate together, worked together, and lived together in the field for over a year, and it has never been my pleasure who was "Duke's" equal as a friend.*
>
> *We can all change our sorrow and grief to admiration that he went as a soldier should go, and my one hope is that when my time comes I may face my God as I know he faced his—like a soldier. I can truthfully say that I would much rather it had been I than he—such men are not often born.*

Howard Van Aradale plays "Taps" at the foot of Captain Mabry's casket during the public memorial service. *Air Combat Command History Office, Joint Base Langley-Eustis.*

Captain Dale Mabry is laid to rest in Arlington National Cemetery on February 25, 1922. *Hampton History Museum 1987-18-105.*

Sergeant Virgil Hoffman is laid to rest in his family's mausoleum in Eaton Rapids, Michigan. *Hampton History Museum 1987-18-107.*

Please accept my love, friendship and understanding. We have both lost something it is not possible to replace.[388]

Stella Hoover couldn't bring herself to travel to Eaton Rapids for the funeral of her beloved Virgil. He was buried on a day when snow blanketed the ground.[389] Like the rest of the victims, local government and civic organizations in Hampton Roads sent grand displays of flowers and notes of their own grief over Virgil's loss. Businesses closed for the day, and the population of Eaton Rapids converged to pay their final respects to the fallen hero and embrace his heartbroken family.

Sergeant Virgil Hoffman was laid to rest in his family's mausoleum as snow fell like frozen tears from the heavens.[390] Virgil's older brother, Earl, wrote a letter to Stella—a woman he had never met but toward whom he felt a kinship. He described Virgil's service in detail and expressed his own sorrow for what Stella must have been feeling. He invited her to travel to Eaton Rapids and meet the Hoffmans, who wanted to accept her as one of their own.[391] Earl's letter brought a small measure of relief to the heartbroken young woman, who vowed that she would soon travel to meet them and say her own, private goodbye to her lost love.

CHAPTER 17

THE BURIAL

Such ships accomplish nothing but to kill people…They are of no value in war.
—Chairman Madden of the House Appropriations Committee[392]

Once funerals ended and the church bells ceased to toll, the debate was brought before Congress over the use of hydrogen in airships or, moreover, Congress's failure to appropriate the funds to transport helium from Fort Worth to Langley for *ROMA*. Clifford Tinker was present, though never asked to speak at open sessions. Top members of the U.S. Army Air Service and the U.S. Navy were quick to jump to Congress's defense. Major General Mason Patrick reported to newspapers, "Blame does not rest with Congress…Statements that failure of Congress to appropriate funds for helium caused the accident are not based on fact. Congress appropriated liberally for development of the process of extracting helium during the war, and has since given all funds requested for further experimental work looking to perfecting the processes."[393]

The debate raged heavy from all members of Congress, debating over semantics, the fine details of the disaster and, of course, the overarching question of fiscal responsibility. There was a clear divide between those who felt that hydrogen's use was still perfectly safe and others who called for the termination of all lighter-than-air programs in the United States. Capitol Hill was hot with talks of hydrogen and the calamity that fell when *ROMA* crashed to the Earth. "Such ships accomplish nothing but to kill people! They are of no value to war!" Chairman Martin B. Madden of the House Appropriations Committee argued.[394]

Congressman Julius Kahn, chairman of the House Military Committee, replied, "The chances are that efforts will be made to prevent a reoccurrence of such a calamity. I feel, however, that so long as the other nations of the world continue experiments regarding flights in the air, our own country will have to keep up with the rest of the world in that regard."[395] Quotations from various sources ran heavy through the newspapers as the investigation continued, enveloped in secrecy and eventually classified.

Lieutenant Colonel Alessandro Guidoni of Italy announced his own conclusions, absolving the Italian government and the airship manufacturers of blame in the tragedy. "Breaking of the control cable of the semi-rigid *ROMA* was responsible for the disaster only because the vessel was flying too low. The *ROMA* courted catastrophe in skirting the Earth. The danger in flying too low was never more forcefully demonstrated than in the sad fate of the *ROMA*, the loss of the precious human freight is felt most keenly by my government and my people."[396] Americans were quick to rush to the defense of the crew, particularly Captain Dale Mabry.

Umberto Nobile issued a statement saying, "It was an unlucky circumstance that the airship met this high-tension wire. Otherwise nothing would ever have destroyed it."[397]

The U.S. Army Air Service began petitioning Congress for funding to replace *ROMA* with another large airship. Much to its dismay, Secretary of War John Weeks made the decision that there wasn't sufficient value of large airships to the Air Service to warrant a replacement for *ROMA*.[398]

Without any true statement released to the public, Major General Patrick would only concede, "The *ROMA* disaster would have happened just the same had the ship been filled with helium instead of hydrogen. It is true, however, that in all probability the loss of life might not have been so great."[399] This reminded Clifford Tinker of the conversations he had with Major Van Nostrand and later Representative Hicks over the lack of funding the army had to transport helium from Fort Worth to Langley Field.

With public anger fueled over the use of hydrogen, the role Congress and the War Department played in failing to appropriate funding to transport the helium for the flight was downplayed in official statements released to the media. This was done as a ploy to preserve the allure of lighter-than-aircraft and maintain funding for these programs. An unnamed official went to the newspapers and said, "I doubt whether the gas policy of the Air Service will be changed. You must remember that the hydrogen in the *ROMA* did not catch fire until the bag was ignited by the broken high tension electric wires. Helium gas, being non-inflammable, is highly desirable, but the present

Sergeant Virgil Hoffman's casket at his funeral in Eaton Rapids, Michigan. *Hampton History Museum 1987-18-108.*

almost prohibiting cost of manufacture, seems to keep it out of our reaching for the time being."[400] Upon reading this statement, Clifford Tinker was taken aback because the fiscal value of helium was already proven with the flights of *C-7* in December.

Neither Master Sergeant Harry Chapman nor Charles Dworack had yet to be stabilized enough to transfer to the hospital at Fortress Monroe. The rest of the survivors were photographed for newspapers and dodged questions from eager reporters. Sergeant Joe Biedenbach and Corporal Alberto Flores were lured out onto the front steps of the hospital to have their photograph taken by post photographer G.W. Lord. None too keen, a smile could only be coaxed out of the two men by their attending nurses.[401]

Debate also began as to what to do with the ruins of *ROMA*. After some persuasion on the part of the investigating team at Langley, it was decided that the battered hulk would be brought back to the field and reassembled in a rudimentary forensic investigation. Lieutenant Byron T. Burt Jr. readily volunteered to oversee the efforts. The wreckage was carefully removed, piece by piece, from the railroad tracks at the Army Quartermaster Depot and transported by barge across the water. Once it arrived at Langley

Field, efforts were made to straighten the twisted, bent and broken pieces of *ROMA*'s remains. They reassembled the keel as best possible, laying it upside down on the ground. Wooden beams were brought in to hold the fragile metal pieces upright. Scraps not pertinent to the investigation were tossed into a pile off to the side of the wreckage. Burt carefully examined the elevator control cables. Sure enough, the cable had snapped,[402] corroborating Captain Reed's theory of the crash. Lieutenant Burt reported his findings to his superiors, and what was left of the great ship was sent to a warehouse for waste material and later sold as scrap.[403]

Back in Washington, D.C., debates over the use of hydrogen continued to rage. In an effort to preserve the use of the gas, the Aeronautic Chamber of Commerce of America issued a callous statement saying:

> *The* ROMA *accident should be regarded the same as an accident in a laboratory. The loss of the* ROMA *should not be regarded as a reflection on the operating personnel or the judgment of the War Department in having purchased it from Italy. It was the obvious thing to do because we have not developed lighter-than-aircraft in this country to the extent that other governments have developed it with public funds. The potentiality of the airship is such that it is bound to be developed, despite an accident or a series of accidents. It is so important and its promise is so far reaching in transportation and national defense that it must continue. We do not abandon car laboratories, mines, railroads or steamships when accidents befall them.[404]*

These insensitive remarks reignited a wave of anger throughout the country. President Warren Harding met with his cabinet, and Secretary of the Interior Albert B. Fail presented a case filled with irrefutable data on the safety of helium, including the data collected from the flights of *C-7*. A tentative bill was also presented that would provide for an increase in procurement, production and preservation of helium.[405]

This bill was highly debated in Congress. Congressman William Raymond Green of Iowa made a motion to completely halt all procurement of helium. He stood before the House of Representatives and said:

> *It may seem somewhat surprising that I make this motion, when I state that Major Thornell was the son of a very old and dear friend of mine....No one save a near relative could more regret the tragic death of Major Thornell than myself. He was a young officer of exceedingly great promise, and his*

death is an unquestionable loss to the Air Service. Mr. Chairman, when the ROMA *went down, as usual we heard a chorus from the papers all over the country that this disaster was the fault of Congress for not having appropriated more money for helium gas…I want to call to the attention of the committee to the fact that so far as I am aware all of the countries that are principally interested in aviation at this time for military purposes have discarded the giant airships.*[406]

It was with this that he urged Congress to consider developing heavier-than-air technology in the place of airships. However, it was concluded that later that year, the helium plant in Fort Worth would be reopened and the use of hydrogen in American airships would be permanently discontinued. Because of *ROMA*, no other man on a United States airship would die because of hydrogen.

June 30, 1922, was the conclusion of that fiscal year. On that day, the War Department returned a budget surplus of $6,498,450.90 to the Department of Treasury. Of this balance, $1,246,865.33 was supposed to have been used for the purposes of transporting supplies for the U.S. Army Air Service. As Clifford Tinker had pointed out in the time prior to the disaster, it would have only cost the War Department $14,000 to transport the helium needed for *ROMA* for her fateful flight on February 21, 1922.

After discovering this in 1925, an enraged Clifford Tinker published an article about his newfound evidence in *Collier's Weekly*. The scathing exposé led him to be called before the Sixty-eighth Congress.[407] In that hearing, evidence came to light that neither Major General Mason Patrick nor the War Department had ever asked Congress for the $14,000 needed to transport the helium from Fort Worth; there would have been no need to with such a large budget surplus. Despite the budget surplus and irrefutable evidence of the efficacy of helium, they negligently chose to use hydrogen for a cost of what would have amounted to approximately $412 per life lost that fateful day.[408]

Chapter 18

EATON RAPIDS, MICHIGAN

It was the opening of that deep, deep wound in my heart which has never healed.
—Stella Hoover in a letter to her mother[409]

In the weeks following the loss of *ROMA*, Stella Hoover began receiving letters from the Hoffman family. Virgil's brother, Earl, and Earl's wife, Nellie, would refer to Stella as "sister" and express their deep love and appreciation for this young woman whom they had never met.[410] To the Hoffman family, she was all they had left of Virgil. Stella returned correspondence, feeling an equal affection for the Hoffman family.

Virgil and Stella were supposed to have been married that July and planned to travel back to Eaton Rapids for Stella to meet his beloved family. Instead, she boarded a train to Michigan with only her memories of him beside her.

Stella spent most of the long ride writing letters to her mother and friends in Phoebus, describing the various people she encountered and the things she saw. Upon arriving in Eaton Rapids, she was met with warm embraces by Virgil's family, including his mother, whom Stella referred to as "Mother Florence."[411] Without hesitation, they readily took her to the cemetery. Unlike the day Virgil was buried, the ground was lush and green and the sky a vibrant blue. Stella followed Earl warily through the cemetery, Nellie clutching her shaking hand. They walked into the large marble family mausoleum, and Earl gestured toward a large brass plaque. It read:

Virgil C. Hoffman
Son of
Benjamin and Florence Hoffman
Born 1900 in Creston, Ohio

Killed in the ROMA Accident
February 21, 1922

Stella felt her body become limp as she fell to her knees. She fiercely wept from the incredible pain she felt throughout her soul. Virgil was truly gone. Earl knelt beside her, choking back his own tears as he placed his arm around her shoulders.

That night, Stella penned a letter to her mother in Phoebus:

> *I arrived in Eaton Rapids all ok. Mother Florence is a dear. Earl was the greatest comfort, his presence felt just like wings. He talked to me just as sweet and nice as you would mother dear.*

> *I visited the place where my lost darling is buried. It certainly is beautiful.*

> *Mother, I had a nervous spell and cried something terrible. It was the opening of that deep, deep wound in my heart that has never healed.*[412]

EPILOGUE

In May 2014, I sought from the Virginia Port Authority (VPA) access to the private property where a monument to *ROMA* and her crew now stands. It was a warm, sunny morning when I met my escort from VPA as he unlocked the chained gate to allow me onto the grounds. The monument is a small, rectangular slab of marble with words slightly etched and painted on the top. A brass plaque is screwed to the front, long lost of its original luster and with words so small that you have to be right against it to be able to read the inscription.

In the years following the disaster, the Army Quartermaster Depot was closed and later became Norfolk International Terminals, a privately owned port for the import and export of goods in and out of the United States. In 1952, dock workers pooled their money together, and the following year, a small dedication service was held. Alberto "Little" Flores was the only *ROMA* survivor in attendance.[413] While the efforts of those workers should not be taken in vain, the monument does not resemble the grand memorial that Virginia Attorney General Colonel John R. Saunders promised in his eulogy during the public memorial service held on February 24, 1922.

Today, the monument is hidden between two bushes, standing at least a quarter mile from where the ship actually went down, on private property belonging to the terminals. Due to safety concerns of the bustling port, the area in which *ROMA* crashed is not accessible to anyone who does not work there. While kneeling in front of the monument, I felt a sense of emptiness and hurt consume me for the forty-five men who went up on *ROMA* that day and the thirty-four who made the ultimate sacrifice.

With more accidents of helium-filled airships in the coming years, military lighter-than-air programs were nearly abandoned in favor of continuing to develop heavier-than-air technologies. The hangar at Langley that once housed *ROMA* was disassembled in 1947,[414] and the nearby hydrogen processing plant was left abandoned. Military housing and offices were built where the hangar once stood. Many of the streets were named to honor the victims of *ROMA*. Joint Base Langley-Eustis is still a vibrant, thriving post where jets fly in the skies where *ROMA* once ascended, and the area where she was assembled is still affectionately referred to as the "LTA Section" to honor the memory of the balloon men who once called these hallowed grounds their home away from home.

For the eleven survivors, they sought no fame or profit from their story. They moved on with their lives as best they could. Charles Dworack spent considerable time in the hospital in the aftermath of the disaster. He permanently suffered from his injuries, compromising his ability to continue to perform his job. The mental anguish he suffered from would perhaps today be considered post-traumatic stress disorder.[415] It took twelve years of petitioning Congress in order to receive any sort of financial compensation for what he had endured.[416]

Master Sergeant Harry Chapman was transferred to Walter Reed Hospital in Washington, D.C., and spent close to a year recovering from his injuries. He was undoubtedly a hero for having returned to the fire to save four of his shipmates' lives. In 1929, he was honored as the first recipient of the Cheney Award for his bravery.[417] Master Sergeant Chapman was the only *ROMA* crew member to receive any sort of commendation for that day. He joined the crew of the navy zeppelin USS *AKRON* before being transferred back to Langley Field in March 1933.[418] Just four weeks later, USS *AKRON* was tragically lost in the Atlantic Ocean, killing seventy-three of the seventy-six passengers and crew members on board. After continuously reenlisting in the U.S. Army Air Service, he fell ill in 1938.[419] On April 20, 1940, Master Sergeant Harry A. Chapman passed away at Fitzsimons Army Hospital in Aurora, Colorado,[420] due to the long-term effects of the injuries he sustained from the *ROMA* disaster.[421]

Private Virden Peek finished out his tenure in the military, eventually marrying and settling down in Indiana. He became active in both his church and the local chapter of the American Legion while working in security and later as a custodian.[422] Ray Hurley returned to McCook Field and resumed his work.[423] It does not appear that the friendship they forged in the fire of *ROMA* that day endured.

Major John Reardon went on to obtain the rank of colonel in the United States Air Force before retiring.[424] He settled in North Carolina and, in 1959, obtained official records regarding the ill-fated ship and his fallen crewmates.[425] In 1960, he penned a fourteen-page essay of his experience on board *ROMA* that day in 1922. Even thirty-eight years later, the images and feelings were still incredibly vibrant and haunting to him. This essay is now on file with the Air Force Historical Research Agency at Maxwell Air Force Base—Colonel Reardon's lasting tribute to his fallen brethren.

Sergeant Joseph Biedenbach also remained in the air service, serving as part of the crew of smaller airships. He married Augustine Boehs, and together they had several children.[426] Joe Biedenbach passed away on June 2, 1957.[427] Just a few short weeks later, his son, First Lieutenant Donald Biedenbach, a pilot in the U.S. Air Force, tragically perished in a training accident when his burning plane crashed into a home in Illinois.[428]

Corporal Alberto Flores also stayed in the air service, eventually retiring from the reserves as a master sergeant. He lived in Belleville, Illinois, where he had many friendships with locals who became a surrogate family to him, though he often visited his own family back home in Puerto Rico. His loved ones remembered him as a fun-loving man who, to his landlord's daughter, Judie, became "Uncle Duke." He loved to regale others with tales of his time in the air service and about serving on the great balloons, though he would never again return to the sky after *ROMA*.[429] Master Sergeant Alberto "Little" Flores was the last living survivor before dying on June 8, 1988, at ninety years old.[430]

Walter McNair returned to the Bureau of Standards in Washington, D.C., continuing his work in researching aeronautical physics, and seemed to fall into obscurity in the passing of time.

Stella Hoover continued to live in Hampton, eventually marrying Amsy Saunders. Together they had many children who were all told stories of Virgil Hoffman. Stella held on to a framed portrait of Virgil, which she kept alongside his silver cigarette case and matchbox. The cigarettes remained inside, untouched by time. When she died in 1949,[431] these mementoes were passed along to her daughter, who lovingly displayed them for historian John B. Mitchell in 1972.[432]

Lieutenants Byron T. Burt Jr. and Clarence H. Welch and Captain Walter J. Reed all continued to have vibrant military careers. All three would eventually earn their wings as pilots of heavier-than-air crafts and serve

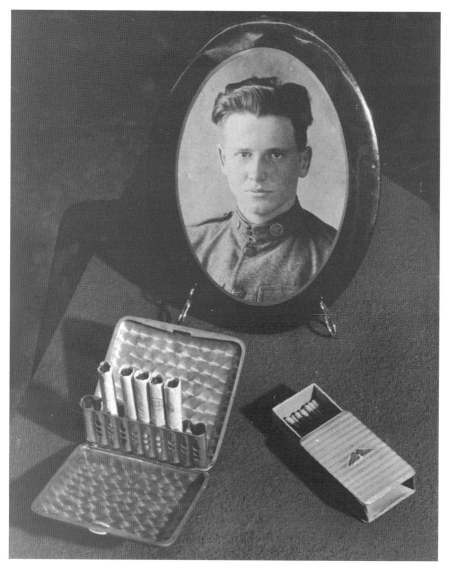

Sergeant Virgil Hoffman's cigarettes and silver cigarette case. *Hampton History Museum 1987-18-126.*

during World War II. Colonel Byron T. Burt Jr. passed away on February 9, 1969, and is fittingly buried within walking distance of his fallen shipmates at Arlington National Cemetery.[433]

Captain Reed was transferred to Lakehurst, New Jersey, where he helped oversee the construction of the ill-fated airship USS *SHENANDOAH*. He and Maria continued to have a very happy marriage, living every day to the fullest extent possible. Along with their two children, they moved all over the world, befriending pioneers of modern military aviation like Hap Arnold and Eddie Rickenbacker. They remained close to Marie Thornell, as well as Dale Mabry's family, finding time to visit with them and speak about days gone by. Walter Reed retired from the United States Air Force as a brigadier general and eventually settled in the U.S. Virgin Islands. In 1963,[434] he passed away, followed by Maria in 1976.[435] They are buried a short distance from Joint Base Langley-Eustis at St. John's Episcopal Church in Hampton, Virginia, within minutes of the house where Maria was raised.

In the late summer of 2014, I had the privilege of speaking to their son, Lieutenant Colonel Walter J. Reed Jr. (USAF, ret.). He told me about his father, and the admiration in his voice was just as infectious as it was palpable. He shared with me about his father's loving, kind personality and how he had been described by friends as the "sweetest man they ever met."[436] When I asked Lieutenant Colonel Reed Jr. if his father ever spoke to him about *ROMA*, he shared with me a story that he had read in his father's diary. After retiring from the U.S. Air Force, the Reeds were traveling back to Hampton aboard an army transport vessel. Upon entering the mouth of the river, Brigadier General Reed looked to the sky and all he could say was that "he had never forgotten."[437]

After tears dried and the victims were laid to rest, *ROMA* fell into the obscurity of the back pages of newspapers before eventually disappearing altogether. There was never a grand monument with the victims' names erected, their

The memorial stone in its current position behind a locked gate at Norfolk International Terminals. *Nancy E. Sheppard.*

story long since forgotten and their lives never recognized for the sacrifice they made in the line of duty. This stain on the War Department and the U.S. Army Air Service was buried, and all that would be left of the thirty-four men who gave their lives on February 21, 1922, would be a marble slab hidden and locked behind a fence.

I will never forget how I felt touching the memorial stone that day. It was as though I was surrounded by the forty-five men who boarded *ROMA* that cold morning in February 1922—forty-five men who were brought together at a single point in time, destined to make history but, much like the memorial stone, overlooked and forgotten as time marched on. No sacrifice so great should ever be forgotten, nor should the story of the great airship *ROMA* and her valiant crew.

WHY DID *ROMA* CRASH?

I have been asked time and time again why *ROMA* crashed. For years, the answer to this question remained shrouded in mystery. However, the truth behind *ROMA*'s failure is no longer a secret.

ROMA was a semi-rigid airship that depended on the pressure within her eleven gas ballonets to maintain structural integrity. She had an articulated keel that was flexible to allow for variations in her non-rigid envelope. The elevator controls ran through hollow tubing from the control cabin, located in the center of the ship (beneath gas compartment 4), to the rudder.

February 21, 1922, was a perfect storm of events. It was common knowledge that the envelope was in poor condition and gas compartment 1 was in terrible shape. While being pulled out of the hangar, a large patch fell off of the top of the ship and landed at John Tucker's feet. It can be surmised that this patch came from the front of the envelope, over the top of gas compartment (ballonet) 1. Also it is important to note that when *ROMA* went up that afternoon, the crew didn't compensate for the weight disparity between the new, lighter Liberty engines and the Ansaldos (which was a difference of approximately 1,200 pounds). Also, *ROMA* was carrying less ballast than usual that day, hence why, on lift off, the tail section of the ship shot up. However, due to the loss of the patch on top of the envelope, gas compartment 1 was unable to maintain hydrogen pressure. With Master Sergeant Chapman unable to valve gas to the most forward ballonet, *ROMA* was nose heavy.

During the flight, the ship's altitude was a paltry five hundred feet. As notated on Mr. McNair's device, she was moving at a speed of seventy-five

ROMA's crew inspects and repairs gas compartment 1 from inside the envelope. *National Museum of the U.S. Air Force.*

ROMA's ground crew prepares to release the ropes holding the ship to the ground before one of her trial flights. *Hampton History Museum 1987-18-56.*

miles per hour—far faster than *ROMA* had ever flown before. At this time, Corporal Alberto Flores observed the front of the ship losing pressure. Due to the low altitude and fast speed, increased pressure was quickly forcing hydrogen through the hole in gas compartment 1.

This put incredible strain on the articulated keel, which was supporting the heavy copula without the assistance of the envelope. Because of the weight of the copula, *ROMA*'s keel began to buckle. The elevator controls were not made of flexible material, and the more the keel buckled, the more strain was put on the controls until the cables snapped. Because the control cables also stabilized the rudder's altitude, the rudder rotated in such a way as to impact the lower vertical stabilizer and become stuck at an approximately forty-five-degree angle. At this time, Lieutenant Byron T. Burt Jr. noticed that the controls would no longer respond.

Due to the low altitude and lack of central control to empty ballast, the crew was unable to keep the ship aloft. And thus she began a powered descent to her inevitable destruction below. The fact remains that, no matter what gas would have been used, *ROMA* was doomed.

Disaster was met when the ship crashed to the ground and the rudder came to rest on a high voltage wire. This sparked the already unstable hydrogen, causing a conflagration that also consumed the gas tanks. It is key

The skeletal frame of *ROMA*'s box kite rudder rests on the high voltage lines that sparked the conflagration. *J. Sargeant Memorial Collection, Norfolk Public Library.*

to remember that the gas tanks were in very poor condition prior to the flight. The resultant explosion and subsequent fire was likely the cause of the majority of the crew members' deaths. If the men did not get out of the wreckage prior to the gas tanks exploding, then they had no chance of escape. As Lieutenant Clarence Welch pointed out, had there been a non-inflammable gas in *ROMA*, there would have been a far better chance of survival because contact with a high-voltage wire would likely not have resulted in the ignition of the gasoline tanks.

It's important to remember, though, that the War Department was fully briefed on the dangers of hydrogen as well as the fiscal advantages of helium. Despite having an abundance of funding to transport helium from Fort Worth, Texas, to Langley Field (as Mr. Tinker proposed), they negligently chose to use hydrogen anyway as a cheaper, more readily available option. Once it was truly realized that hydrogen was the causality behind the deaths of thirty-three of the thirty-four men who died with *ROMA*, the story was buried and the War Department never accepted responsibility for its negligence.

The only mystery that remains is why the forward set of motors was not shut off. It is likely that, due to the structural failure of the ship and the rapid descent that she endured, the message to idle the motors never reached the forward engineers.

ACKNOWLEDGEMENTS

I want to give special thanks to the following people and agencies for their generous help and support throughout this book. I am humbled, beyond words, for all of the time, experience and dedication you gave. Without you, this would have never been possible!

To my husband, Joshua: Thank you for everything you have given me: your time, patience, support and love. Thank you for being my best friend, advisor, editor and research assistant throughout this process. I love you more than words can say.

To my beautiful children, Emory and Ben: Thank you for your time, patience, understanding and being the most wonderful children a mother could ask for. You are both my heroes, and there is no measure of how proud I am of you. I love you both to the ends of the universe.

The three of you mean everything to me!

A special thank you to my parents, Captain and Mrs. James R. Miller (USN, ret.). Thank you for always supporting, loving and believing in me. Mom, thank you for your tireless friendship, and Dad, you're the finest history teacher I've ever had. I profoundly admire and love you both. I also would like to thank my sister, Jennifer Miller, for inspiring me to start writing many years ago.

This book would never have been possible without the support of the loved ones connected to the crew of *ROMA*. Thank you to Lieutenant Colonel Walter J. Reed Jr. (USAF, ret.), Judie Louden, Jim Saunders, Miguel Navas and the Flores family, Richard Treadway, Eric Howdershelt and Graydon Gorby.

Acknowledgements

I am indebted to the following historians and institutions for their efforts, support, imagery, collections and resources: Bethany Austin and the staff of the Hampton History Museum, Mike Dugre and the staff of Air Combat Command History Office (Joint Base Langley–Eustis), Troy Valos and the J. Sargeant Memorial Collection (Norfolk Public Library), Mrs. Lynn Gamma and the staff of the Air Force Historical Research Agency (Maxwell Air Force Base), United States Air Force History Office (Pentagon), John B. Mitchell, David Bragg, Robert Hitchings, Fabio Iaconianni (freelance photographer and blogger), Jennifer Doerr, Professor Antonio Ventre, Avia-it, Bill Welker, Lieutenant Stevenson of the Virginia Port Authority, John T. Ball, University of Akron, Sidney Public Library (Sidney, Iowa), Raffaelle Migneco, Amici di Umberto Nobile, Mike Perkins of the Indianapolis Public Library, Louisville Free Public Library, Amy Vozar, Diana Keto-Lambert, the Lighter-Than-Air Society and Northeast Ohio Blimp Spotters.

Lastly, thank you to J. Banks Smither and the team at The History Press.

Thank you for helping me honor these brave men.

NOTES

CHAPTER 1

1. Kenneth L. Roberts, "Italy From a Dirigible Window," *Saturday Evening Post*, August 13, 1921, 70.
2. "Lighter-than-air technology" refers to dirigibles, including blimps and zeppelins.
3. John B. Mitchell, *The Army Airship* ROMA (Hampton, VA: Syms-Eaton Museum, 1973), 9.
4. Ibid., 9.
5. Ibid, 9.
6. Majors Davenport Johnson, John H. Jouett and Joseph McNarney, *Proceedings of a Board of Officers in Regard to the Accident to the Airship ROMA, Feb. 21, 1922. Langley Field, VA* (Langley Field, VA: United States Army Air Service, 1922), 66.
7. "Dirigible '*ROMA*' Volo di Consegna All Equipaggio Americano, Roma-Napoli, 15 Marzo 1921," Stabilimento di Costruzoni, Aeronautiche Laboratorio Fotografico, 1921.
8. Ibid.
9. Ibid.
10. Mitchell, *Army Airship ROMA*, 9.
11. Ibid.
12. Ibid.
13. Ibid.
14. Ibid.

15. Ibid.

16. A handling crew was the crew on the ground in charge of holding, releasing and catching rigging cables connected to an airship.

Chapter 2

17. "Envoy Johnson Takes 500-Mile Air Trip." *New York Times*, March 16, 1921, 9.

18. "Many Thousand Warplanes Are Store [*sic*]—Useless" *Muskogee County Democrat*, August 18, 1921, 1.

19. Jordan Golson, "WWI Zeppelins: Not Too Deadly, but Scary as Hell," Wired Online, last modified October 3, 2014, http://www.wired.com/2014/10/world-war-i-zeppelins/.

20. "World War I: Treaties and Reparations," United States Holocaust Memorial Museum, accessed August 4, 2015, http://www.ushmm.org/wlc/en/article.php?ModuleId=10007428.

21. First Lieutenant Robert S. Olmsted, "How the Army Will Use the Airship *ROMA*." *U.S. Air Service*, January 1922, 17–20.

22. Roberts, "Italy From a Dirigible Window," 1.

23. "Pony balloons" were small blimps capable of carrying only a very small number of passengers.

24. Roberts, "Italy From a Dirigible Window," 4.

25. Ibid.

26. Ibid., 70.

27. Ibid., 63.

28. Johnson, Jouett and McNarney, "Proceedings," 9.

29. Roberts, "Italy From a Dirigible Window," 69.

30. "Airship Makes Speedy Voyage." *Ogden Standard Examiner*, 16 March 1921, 2.

31. Roberts, "Italy From a Dirigible Window," 70.

32. Ibid.

33. *New York Times*, "Envoy Johnson," 9.

Chapter 3

34. *Brooklyn Daily Eagle*, "Airship *ROMA* Due to Reach Here Today," June 2, 1921, 2.

35. Johnson, Jouett and McNarney, "Proceedings," 45.

36. *Harrisburg (PA) Evening News*, "King Emmanuel Sees Vatican From Air," October 28, 1920, 2.

37. Aerocave "*ROMA*"—Italian Flight Log, 1.

38. "Rome to Rio de Janeiro," *International Shipping Digest* 1, no. 1 (1919): 51.

39. An envelope on an airship refers to the outer layer of fabric of the balloon.

40. Curtis, Copp and Mitchell, *Langley Field*, 91.

41. "Ballonets" were mini balloons within the envelope that contained the lifting gas. Surrounding each ballonet were pockets for air.

42. "The New Italian Usuelli Semi-Rigid Airship," 44.

43. Grossman, "Hindenburg Flight Operations and Procedures."

44. "Report on Accident to the Airship *ROMA*: Official War Department Report Fails to Determine with Absolute Accuracy the Causes of the Accident." *Aviation*, August 7, 1922: 148–52.

45. A. Black and D.R. Black, "Italians Adapt Semi-Rigid Construction to Large Dirigible." *Automotive Industries* 43 (1920): 758–59.

46. *Flight: The Aircraft Engineer & Airships*, "The New Italian Usuelli Semi-Rigid Airship." September 20, 1920, 44.

47. Lieutenant Walter J. Reed, *Aircraft Log Book, Ansaldo* (Langley Field, VA: U.S. Army Air Service, 1921), 1.

48. Ibid.

49. "The New Italian Usuelli Semi-Rigid Airship," 44.

50 *Contract No. 7105-B, Air Service Order No. 900023-B* (Air Service, United States Army, 1921).

51. *Brooklyn Daily Eagle*, "Airship *ROMA* Due to Reach Here Today," June 2, 1921, 3.

52. Johnson, Jouett and McNarney, "Proceedings," 74.

53. *Kansas City (MO) Star*, "Prepare *ROMA* for Shipment," April 30, 1921, 31.

54. Major Percy E. Van Nostrand, "Lessons Learned from the '*ROMA*' Accident," *U.S. Air Services* 7, no. 1 (1922): 16, 28.

55. Johnson, Jouett and McNarney, "Proceedings," 66.

56. Ibid., 31.

57. "Shipmaster" or "coxswain" on a U.S. Army Air Service airship was the most senior enlisted member of the crew.

58. Ibid., 65.

59. Mitchell, *Army Airship ROMA*, 12.

60. "1862: Battle of the Ironclads." *This Day in History*, History.org, http://www.history.com/this-day-in-history/battle-of-the-ironclads.

61. "The Fort Monroe Story: 400 Years of History." *Fort Monroe: Where Freedom Lives*, http://www.fmauthority.com/about/fort-monroe/history.

62. Hampton Flats is the body of water separating the James River from the Elizabeth River.

63. "Enduring Legacy: The Jamestown 1907 Exposition and the U.S. Navy." Hampton Roads Naval Museum, http://www.history.navy.mil/museums/hrnm/interactive/1907exposition/index.htm.

64. "Langley History." NASA.gov, https://www.nasa.gov/centers/langley/about/history.html#.Vb92TvlViko.

65. Ibid.

66. Office of History, *Langley: 1916–2003*, 3.

67. Ibid.

68. *Washington Herald*, "The Army's New Airship, *ROMA*, Nearly Rivals Lost *ZR-2*," August 30, 1921, 4.

69. *Crew Members and Navigation Captain during Flight of Airship ZDUS-1 Flies over Sky of Norfolk, Virginia.* Featuring Captain Dale Mabry and members of the U.S. Army Air Service. USA: CriticalPast.com. Film. http://www.criticalpast.com/video/65675036617_airship-ZDUS-1_crew-members_ships-at-harbor_navigation-captain

70. Earl Hoffman, letter to Stella Hoover.

71. Mitchell, *Army Airship ROMA*, 15.

72. "Naval and Military Aeronautics." *Aerial Age Weekly* 13, July 18, 1921, 447.

CHAPTER 4

73. *Washington Herald*, "The Army's New Airship, *ROMA*, Nearly Rivals Lost ZR-2," August 30, 1921, 4.

74. Mitchell, *Army Airship ROMA*, 12.

75. *San Francisco Chronicle*, "Survivor of *ROMA* Has Lucky Star," February 22, 1922,

76. Mitchell, *Army Airship ROMA*, 20.

77. Johnson, Jouett and McNarney, "Proceedings," 8.

78. Ibid., 123.

79. Jeremy Plester, "Weatherwatch: Fog Led to destruction of 1921 airship." TheGuardian.com, August 18, 2011, http://www.theguardian.com/news/2011/aug/18/weatherwatch-airship-fog-zeppelin.

80. "R38/ZRII." The Airship Heritage Trust, http://www.airshipsonline.com/airships/r38.

81. Inquiry into Operations of the United States Air Services: Hearing Before the Select Committee of Inquiry Into Operations of the United States Air Services (United States Congress, Washington, D.C.: Sixty-eighth Congress, 1925), 208.

82. Winters, "America's Latest Airship," 259–60.

83. Johnson, Jouett and McNarney, "Proceedings," 143.

84. "The U.S. Army Dirigible *ROMA*: America's First Major Airship Disaster." *Gasbag Journal*, December 1, 1999, 15.

85. *Brooklyn (NY) Daily Eagle*. "Airship *ROMA* Due to Reach Here Today," 3.

86. Hewitt, "Biggest Navy Blimp Burns," 1,650.

87. Girard Chambers, interview by Jennifer Doerr, 2005.

88. *Log of the* ROMA (Langley Field, VA: U.S. Army Air Service, 1922), 2.

89. Curtis, Copp and Mitchell, *Langley Field: The Early Years*, 90.

90. "*ROMA* Is Ready," *Aerial Age Weekly* 14, no. 1 (1921): 244.

91. Ibid.

92. *Washington Post*, "The Army's Airship," October 30, 1921, 35.

93. "Engines on the Airship *ROMA* to Be Changed." *Aerial Age Weekly*, January 30, 1922, 499.

94. *Cincinnati Enquirer*, "*ROMA* to Visit Akron," October 29, 1921, 4.

95. *Lawrence (KS) Daily Journal-World*, "Regular Airflights," November 28, 1921, 1.

96. *Log of the* ROMA, 10.

CHAPTER 5

97. Mitchell, *Army Airship* ROMA, 1.

98. Ibid.

99. Ibid., 5.

100. Ibid., 1.

101. Mark St. John Erickson, "The World's Largest Lighter-Than-Airship Flew into History at Langley Field," *Daily Press*, November 15, 2013, online.

102. *Log of the* ROMA, 13.

103. The crow's nest on *ROMA* was a hatch on the top of the copula that allowed Corporal Flores to monitor the topside of the envelope and the structural integrity of the copula.

104. St. John Erickson, "The World's Largest Lighter-Than-Airship."

105. The directional pilot was in charge of the forward movement of an airship while the altitude pilot was in charge of any change in elevation of the ship.

106. Grossman, "Hindenburg Flight Operations and Procedures."

107. *Log of the* ROMA, 13.

108. "Engines on the Airship *ROMA* to Be Changed," *Aerial Age Weekly*, January 30, 1922, 499.

109. Ibid., 12.

110. St. John Erickson, "The World's Largest Lighter-Than-Airship."

111. Lieutenant Colonel Walter J. Reed Jr., phone interview, August 2014.

112. Hampton Normal and Agricultural Institute would later become known as Hampton University.

113. Girard Chambers, interview by Jennifer Doerr, 2005.

114. Spooner, "The '*ROMA*' Flies Again," 862.

115. Mitchell, *Army Airship* ROMA, 5.

116. Ibid.

117. St. John Erickson, "The World's Largest Lighter-Than-Airship."

CHAPTER 6

118. Mitchell, *Army Airship* ROMA, 15.

119. Ibid., 5.

120. The Bureau of Standards was the predecessor to the National Institute of Standards and Technology.

121. *Log of the* ROMA, 13.

122. *Indianapolis (IN) Star*, "Member of Graveyard Club, Made Up of ROMA's Crew, Predicted Death to Girl," February 24, 1.

123. *Log of the* ROMA, 15.

124. "The C-7 Tests," *United States Naval Institute Proceedings* 48 (1922), 305.

125. "Inquiry into Operations," 209.

126. *Log of the* ROMA, 19.

127. Mitchell, *Army Airship* ROMA, 13.

128. Ibid., 14.

129. Ibid.

130. *Washington Post*, "ROMA Battles Gale," December 22, 1921, 1.

131. "Inquiry into Operations," 209.

132. Ibid.

133. "Engines on the Airship," *Aerial Age Weekly*, 499.

134. *Washington Post*, "ROMA Battles Gale," 1.

135. U.S. National Archives, "Lighter Than Air, 1937," YouTube, n.d.https://www.youtube.com/watch?v=7hBZolnvWbQ.

136. "U.S. Army Dirigible *ROMA*," *Gasbag Journal*, 15.

137. Curtis, Copp and Mitchell, *Langley Field*, 92.

138. *Log of the* ROMA, 21.

139. Ibid.

140. *Los Angeles Times*, "ROMA Now Queen of Air," December 22, 1921, 12

141. Ibid.

142. U.S. National Archives, "Lighter Than Air, 1937," YouTube, n.d.https://www.youtube.com/watch?v=7hBZolnvWbQ.

143. Mitchell, *Army Airship* ROMA, 14.

144. Ibid.

145. *Washington Post*, "ROMA Battles Gale," 1.

146. Ibid.

147. "Lighter Than Air, 1937."

148. Ibid.

149. *Log of the* ROMA, 21.

150. Ibid.

151. *Cincinnati Enquirer,* "Statement on *ROMA* Is Reiterated," February 26, 1922.

152. Mitchell, *Army Airship* ROMA, 14.

153. *Log of the* ROMA, 21.

154. Ibid.

155. Ibid.

CHAPTER 7

156. Watts and Hayes, "*ROMA* Airship Disaster Over Norfolk."

157. *Washington Post,* "*ROMA* Survivor and His Mascot," February 26, 1922, 6.

158. Johnson, Jouett and McNarney, "Proceedings," 103.

159. Mitchell, *Army Airship* ROMA, 19.

160. Ibid., 12.

161. Ibid., 15.

162. Ibid.

163. Ibid.

164. Ibid., 12.

165. Johnson, Jouett and McNarney, "Proceedings," 39.

166. Watts and Hayes, "*ROMA* Airship Disaster Over Norfolk."

167. "*ROMA* Tragedy," National Museum of the U.S. Air Force, last modified April 9, 2015, http://www.nationalmuseum.af.mil/Visit/MuseumExhibits/FactSheets/Display/tabid/509/Article/196943/roma-tragedy.aspx.

168. *Log of the* ROMA, 21–26.

169. Dale Mabry, letter, from Air Force Historical Research Agency.

170. Johnson, Jouett and McNarney, "Proceedings," 7.

171. "Inquiry into Operations," 212. In the early 1920s, "non-inflammable" was synonymous with the modern "inflammable," and "non-inflammable" was used in contemporary works, including official government documents and newspaper articles.

172. Ibid.

173. This was a subcommittee of the Military Affairs Committee.

174. "Inquiry into Operations," 210.

175. Ibid.

176. Johnson, Jouett and McNarney, "Proceedings," 9.

177. *New York Times,* "Sad Fabric of *ROMA* Was in 'Bad Shape,'" February 24, 1922, 1.

178. *Indianapolis (IN) Star,* "Member of Graveyard Club," 1.

179. Ibid.

180. Elevator controls were two ropes in the control cabin that were pulled to change the horizontal and vertical surfaces on the rudder.
181. *Log of the* ROMA, 22.
182. Ibid., 22–27.
183. Each Liberty engine weighed approximately two hundred pounds less than its Ansaldo counterpart.
184. Walter Reed Jr., phone interview with the author.
185. *Schenectady (NY) Gazette*, "34 Men Roasted Alive in Airship *ROMA* Crashing on High Power Wire; Only 11 Out of 45 Passengers," February 22, 1922, 1.
186. Johnson, Jouett and McNarney, "Proceedings," 29.
187. Nellie Hoffman, letter to Stella Hoover, 1–5.
188. Mitchell, *Army Airship* ROMA, 19.
189. *Washington Post*, "Dance Celebrated Launching of Fatal Flight of the *ROMA*," February 22, 1922, 11.
190. "Cover" is military vernacular for uniform hat.
191. Johnson, Jouett and McNarney, "Proceedings," 116.
192. Ibid.
193. *Fifty-third Annual Report of West Point*, 64.
194. Reardon, "The Semi-Rigid Airship *ROMA*," 3.
195. Ibid.
196. Ibid.
197. Ibid.
198. Mitchell, *Army Airship* ROMA, 19.

CHAPTER 8

199. Ibid.
200. Ibid.
201. Johnson, Jouett and McNarney, "Proceedings," 29.
202. Reardon, "The Semi-Rigid Airship '*ROMA*,'" 6.
203. Ibid, 5–6.
204. Ibid., 5.
205. Johnson, Jouett and McNarney, "Proceedings," 83.
206. "Report on Accident to the Airship *ROMA*: Official Report Fails to Determine with Absolute Accuracy the Causes of the Accident," *Aviation*, August 1922, 148–52.
207. Ibid., phone interview.
208. Lieutenant Colonel Walter J. Reed Jr., letter to the author, August 2014.
209. Johnson, Jouett and McNarney, "Proceedings," 108.
210. "Lieutenant General William E. Kepner," United States Air Force, http://www.af.mil/AboutUs/Biographies/Display/tabid/225/

Article/106566/lieutenant-general-william-e-kepner.aspx   (accessed January 5, 2015).

211. *A-4* was a typical U.S. Army Air Service airship. This ship was a small blimp that could be manned by only a few crew members. It had a small, rectangular, open-air body, much resembling the airplanes of the day, suspended by wires from a non-rigid balloon.

212. Watts and Hayes, "*ROMA* Airship Disaster Over Norfolk."

213. Johnson, Jouett and McNarney, "Proceedings," 54.

214. Ibid., 80.

215. Rouse, "Doomed Airship Called Langley Home."

216. Corporal Curro was not part of *ROMA*'s crew on February 21, 1922.

217. Mitchell, *Army Airship* ROMA, 19.

218. Reardon, "Semi-Rigid Airship '*ROMA*,'" 6–7.

219. Johnson, Jouett and McNarney, "Proceedings," 9.

220. Reardon, "Semi-Rigid Airship '*ROMA*,'" 8.

221. Ibid.

222. Johnson, Jouett and McNarney, "Proceedings," 5.

223. Watts and Hayes, "*ROMA* Airship Disaster Over Norfolk."

CHAPTER 9

224. *New York Tribune*, "Fight Started for Funds to Develop Helium," February 22, 1922, 2.

225. Reardon, "Semi-Rigid Airship '*ROMA*,'" 7.

226. Johnson, Jouett and McNarney, "Proceedings," 133.

227. Watts and Hayes, "*ROMA* Airship Disaster Over Norfolk."

228. Reardon, "The Semi-Rigid Airship '*ROMA*.'" 8.

229. Johnson, Jouett, and McNarney, "Proceedings," 68.

230. Air scoops on an airship direct the air exhaust into the air ballonets of the ship, allowing the pilot to fill them.

231. Johnson, Jouett and McNarney, "Proceedings," 72, 83.

232. Ibid., 83.

233. Ibid., 68.

234. Watts and Hayes, "*ROMA* Airship Disaster Over Norfolk."

235. Johnson, Jouett and McNarney, "Proceedings," 7.

236. Reardon, "Semi-Rigid Airship '*ROMA*,'" 8.

237. Johnson, Jouett and McNarney, "Proceedings," 117.

238. Ibid., 80.

239. Ibid., 142.

240. Mitchell, *Army Airship* ROMA, 19.

241. Johnson, Jouett and McNarney, "Proceedings," 68.

242. Watts and Hayes, "*ROMA* Airship Disaster Over Norfolk."
243. Johnson, Jouett and McNarney, "Proceedings," 55.
244. Ibid., 35.
245. Rouse, "Doomed Airship."
246. Johnson, Jouett and McNarney, "Proceedings," 10.
247. Ibid., 35.
248. There are two different types of lift for an airship. Static lift is when an airship increases elevation without the use of thrust from their propellers or any other mechanical means. Dynamic lift, on the other hand, requires mechanical help increase elevation.
249. Ibid, 51.
250. Ibid., 35, 55, 65.
251. Ibid., 35.
252. "Fight Started for Funds to Develop Helium," 2.

CHAPTER 10

253. Mitchell, *Army Airship ROMA*, 20.
254. Johnson, Jouett and McNarney, "Proceedings," 78.
255. Ibid., 10.
256. Ibid., 78.
257. Ibid., 83.
258. Mitchell, *Army Airship* ROMA, 19.
259. Reardon, "The Semi-Rigid Airship '*ROMA*,'" 9.
260. Johnson, Jouett and McNarney, "Proceedings," 83.
261. Ibid., 117.
262. Mitchell, *Army Airship* ROMA, 20.
263. Johnson, Jouett and McNarney, "Proceedings," 6.
264. Ibid., 35.
265. Ibid., 9.
266. Ibid., 9.
267. Ibid., 35.
268. Ibid., 9.
269. Ibid., 19.
270. Ibid., 20.
271. Ibid, 36.
272. Watts and Hayes, "The *ROMA* Airship Disaster Over Norfolk."
273. Mitchell, *Army Airship* ROMA, 20.
274. Rouse, "Doomed Airship."
275. Mitchell, *Army Airship* ROMA, 20.
276. Watts and Hayes, "The *ROMA* Airship Disaster Over Norfolk."

277. Johnson, Jouett and McNarney, "Proceedings," 26.

278. Ibid., 117.

279. Ibid., 26.

280. "Fight Started for Funds," 3.

281. Reardon, "The Semi-Rigid Airship *ROMA*," 10.

282. *Oakland Tribune*, "Dirigible ROMA Destroyed by Explosion; Many Dead," February 21, 1922, 3.

283. Reardon, "The Semi-Rigid Airship '*ROMA*,'" 10.

284. Rouse, "Doomed Airship."

285. Johnson, Jouett and McNarney, "Proceedings," 24.

286. Reardon, "The Semi-Rigid Airship '*ROMA*'," 10.

287. Johnson, Jouett and McNarney, "Proceedings," 36.

288. *Sheboygan Press Telegram*, "Tells of Disaster," February 23, 1922, 9.

289. Johnson, Jouett and McNarney, "Proceedings," 11.

290. Rouse, "Doomed Airship."

291. Mitchell, *Army Airship* ROMA, 20.

Chapter 11

292. Rouse, "Doomed Airship."

293. Johnson, Jouett and McNarney, "Proceedings," 24.

294. *New York Times*, "Giant Army Dirigible Wrecked," February 22, 1922, 2.

295. Johnson, Jouett and McNarney, "Proceedings," 24.

296. "Injured Survivors Tell of Disaster," *New York Times*, February 22, 1922, http://query.nytimes.com/mem/archive-free/pdf?res=9B0DEFD71E3 0EE3ABC4A51DFB4668389639EDE (accessed March 13, 2014).

297. Johnson, Jouett and McNarney, "Proceedings," 10–11.

298. Mitchell, *Army Airship* ROMA, 20.

299. Reardon, "The Semi-Rigid Airship '*ROMA*,'" 10.

300. Mitchell, *Army Airship* ROMA, 20.

301. Johnson, Jouett and McNarney, "Proceedings," 26–27.

302. Ibid., 117.

303. Reardon, "The Semi-Rigid Airship '*ROMA*,'" 11.

304. *Fort Wayne (IN) Sentinel*, "Cause Is Mystery," February 22, 1922, 14.

305. Johnson, Jouett and McNarney, "Proceedings," 27.

306 This man was Charles Dworack. When the crash occurred, he was caught between a few gas tanks. He barely escaped before the final explosion. He was burned so terribly that when his rescuers went to wipe the mud from his face, his skin peeled off as well.

307. Ibid., 37.

308. Ibid., 27.

309. Ibid., 54.

310. Ibid., 30–31.

311. Mitchell, *Army Airship* ROMA, 20.

312. Johnson, Jouett and McNarney, "Proceedings," 12.

313. Ibid., 55.

314. Ibid., 54.

315. Ibid, 55.

Chapter 12

316. Ibid., 27.

317. *New York Times*, "Giant Army Dirigible Wrecked," 1.

318. Johnson, Jouett and McNarney, "Proceedings," 34.

319. *Washington Post*, "Board Again Quizzes Survivors of *ROMA*," March 12, 1922, 4.

320. *New York Times*, "Giant Army Dirigible Wrecked," 1.

321. Johnson, Jouett and McNarney, "Proceedings," 27.

322. Ibid., 55.

323. Ibid., 18.

324. Ibid., 84.

325. Reed Jr., phone interview.

326. Reardon, "The Semi-Rigid Airship '*ROMA*'," 12.

327. Johnson, Jouett and McNarney, "Proceedings," 60.

328. *Muskogee (OK) Times-Democrat*, "Inquiry to Fix Cause Is Begun in '*ROMA*' Crash," February 22, 1922, 10.

329. Johnson, Jouett and McNarney, "Proceedings," 12.

330. *New York Times*, "Injured Survivors Tell of Disaster," February 22, 1922, 1.

Chapter 13

331. Lieutenant General William E. Kepner, "LT GEN William E. Kepner (USAF, ret.) in a letter to historian John B. Mitchell," John Mitchell *ROMA* Research Accn. #X.66, Hampton History Museum, Hampton, VA, 1972.

332. Johnson, Jouett and McNarney, "Proceedings," 37.

333. Dorothy Rouse-Bottom, interview by Jennifer Doerr, 2005.

334. Ibid., 89.

335. Ibid., 92.

336. Ibid., 92.

337. Ibid., 93.

338. *Parsons Daily Sun*, "Witness Says *ROMA* Death Result Burns," February 25, 1922, 1.

339. Reardon, "The Semi-Rigid Airship '*ROMA*,'" 13.

340. Johnson, Jouett and McNarney, "Proceedings," 84.

341. Mitchell, *Army Airship ROMA*, 22.

342. Ibid., 22.

343. "Inquiry to Fix Cause Is Begun in '*ROMA*' Crash," 10.

344. Johnson, Jouett and McNarney, "Proceedings," 100.

345. Mitchell, *Army Airship* ROMA, 22.

346. Reardon, "The Semi-Rigid Airship '*ROMA*,'" 13.

347. Mitchell, *Army Airship* ROMA, 22.

CHAPTER 14

348. "Obituaries," *Fifty-Third Annual Report of the Association of Graduates of the Untied States Military Academy at West Point*, June 1922, 89.

349. Mitchell, *Army Airship* ROMA, 22.

350. *Norfolk (VA) Times-Dispatch*, "Cleaning Up Debris of *ROMA* at the Base," February 27, 1922, 1.

351. Johnson, Jouett and McNarney, "Proceedings," 3.

352. Mitchell, *Army Airship* ROMA, 22.

353. *New York Times*, "Lieut. Riley Elated at *ROMA* Assignment," February 22, 1922, 1.

354. "Obituaries," *Fifty-Third Annual Report of the Association of Graduates of the Untied States Military Academy at West Point*, June 1922, 89.

355. Ibid.

356. Lieutenant General William E. Kepner, "LT GEN William E. Kepner (USAF, ret.) in a letter to historian John B. Mitchell," John Mitchell *ROMA* Research Accn. #X.66, Hampton History Museum, Hampton, VA, 1972. In a letter written to historian John B. Mitchell, Lieutenant General Kepner describes telling sweethearts of the dead about their lost men. While this interaction is only surmised, it is safe to say that this is a fair assumption of what would have happened.

357. *El Paso (TX) Herald*, "Germans Send Sympathies," February 23, 1922, 7.

CHAPTER 15

358. Kolbert, *Sheboygan Press Telegram*, 1.

359. Mitchell, *Army Airship* ROMA, 24.

360. Lieutenant General William E. Kepner, "LT GEN William E. Kepner (USAF, ret.) in a letter to historian John B. Mitchell," John

Mitchell *ROMA* Research Accn. #X.66, Hampton History Museum, Hampton, VA, 1972.

361. *Norfolk (VA) Ledger-Dispatch*, "Broken Controls Caused *ROMA* Crash," February 22, 1922, 1.

362. *Harrisburg (PA) Telegraph*, "C.E. Diehl Has Piece of Fabric From *ROMA*," March 15, 1922, 13.

363. Robert Hitchings, personal interview, Norfolk, VA, May 2014..

364. *Norfolk (VA) Ledger-Dispatch*, "Legislatures See Waterfront From Deck of Special Steamer; View Scene of *ROMA* Disaster This Afternoon," February 22, 1922, 2.

365. *Norfolk (VA) Ledger-Dispatch*, "Broken Controls," 1.

366. Ibid.,"Survivor Tells of Airship's Crash," February 22, 1922, 1.

367. Johnson, Jouett and McNarney, "Proceedings," 3.

368. Ibid., 32.

369. *Evening (OH) Review*, "Open Probe of Crash of Dirigible *ROMA*." February 22, 1922, 1.

CHAPTER 16

370. Johnson, Jouett and McNarney, "Proceedings,"26.

371. A letter that was allegedly written by Lieutenant Clifford Smythe was also heavily printed. This letter was presented by a family friend to the media. While it was confirmed by Lieutenant Smythe's brother, his father and Colonel Arthur Fisher contested it. The actual letter was never produced and therefore cannot be confirmed.

372. William K. Hutchinson, *Oakland Tribune*, 1.

373. Mitchell, *Army Airship* ROMA, 25.

374. "Handwritten Notes on *ROMA*," Langley Air Force Base, VA, Air Combat Command, Joint Base Langley-Eustis.

375. Mitchell, *Army Airship* ROMA, 25.

376. *Virginian-Pilot*, "Airman's Rose Falls on Bier of *ROMA* Head," February 25, 1922, 1.

377. According to the U.S. Air Force History Office at the United States Pentagon, Private John E. Thompson's body would have been collected from the church by the Army Quartermaster Corps. He was later buried in an unremarkable grave in his hometown.

378. Mitchell, *Army Airship* ROMA, 25.

379. Ibid., 25.

380. *Richmond (VA) Times-Dispatch*, "Colonel Saunders to Represent Governor at Public Service," February 24, 1922, 1.

381. Mitchell, *Army Airship* ROMA, 26.

382. *Richmond (VA) Times-Dispatch*, "Colonel Saunders to Represent Governor at Public Service," February 24, 1922, 1.

383. Mitchell, *Army Airship* ROMA, 26.

384. Ibid., 27.

385. Ibid., 27.

386. *Norfolk (VA) Times-Dispatch*, "Pay Last Tribute to *ROMA* Victims," February 25, 1922, 1.

387. Ibid.

388. "Obituaries," 65.

389. John Mitchell *ROMA* Research Accn. #X.66, Hampton, VA: Hampton History Museum, n.d.

390. Ibid.

391. Earl Hoffman, letter to Stella Hoover.

CHAPTER 17

392. *Logansport (IN) Pharos-Tribune*, "America Buries *ROMA* Victims," February 23, 1922, 1.

393. *Oakland (CA) Tribune*, "Congress Absolved by Aero Chieftan," February 23, 1922, 1.

394. *Logansport (IN) Pharos-Tribune*, "America Buries *ROMA* Victims," 1.

395. Leo C. Forrester Jr., "Remembering the *ROMA*," Researcher News Online, http://researchernews.larc.nasa.gov/archives/2002/022202/Roma.html (accessed August 10, 2014).

396. *Oakland (CA) Tribune*, "*ROMA* Sailing Too Low, Italian Expert Claims," February 23, 1922, 1.

397. Umberto Nobile, "Zeppelin Design: Umberto Nobile and the Italian N-1 Airship," interview by Kenneth Leish, 1960.

398. *Chicago Daily Tribune*, "No More *ROMA* Type of Ships, Weeks Implies," February 24, 1922, 10.

399. Forrester, "Remembering the *ROMA*."

400. *Sheboygan (WI) Press Telegram*, "Stop Building Dirigibles for U.S. Use," February 23, 1922, 9.

401. *Norfolk (VA) Post*, "'Better, Thank You, Sir!' *ROMA* Injured Say," February 24, 1922, 1.

402. U.S. Army Air Service, "Images of the Reconstruction of *ROMA*," courtesy of Air Force Historical Research Agency, Maxwell Air Force Base, 1922.

403. *Indianapolis (IN) News*, "*ROMA*'s Battered Hulk Back at Langley Field," June 24, 1922, 2.

404. *Sheboygan (WI) Press Telegram*, "Work Must Not Stop," February 23, 1922, 9.

405. *Virginian-Pilot*, "May Urge Helium Production for Use in Dirigibles," February 25, 1922, 1.

406. Sixty-seventh Congress Congressional Record, 4,523.

407. "Inquiry into Operations," 90.

408. Ibid., 212.

CHAPTER 18

409. Stella Hoover, "Stella Hoover to Her Mother," June 1922, courtesy of Air Combat Command History Office, Joint Base Langley-Eustis.

410. Hoffman, "Earl Hoffman in a Letter to Stella Hoover," 1–5.

411. Hoover, "Stella Hoover to Her Mother," 1.

412. Ibid., 1–3.

EPILOGUE

413. John B. Mitchell, "The *ROMA* and Her Crew." *American Aviation Historical Society Journal* 18, no. 1 (Spring 1973): 6.

414. Vic Johnston, "Do You See What I See?" First Fighter Wing Public Affairs, http://www.jble.af.mil/shared/media/document/AFD-091215-069.pdf (accessed August 12, 2013).

415. Charles W. Dworack: Report to Accompany H.R. 666, Washington, D.C., Seventy-third Congress, Second Session, 1934, 3.

416. Ibid., 1.

417. *Washington Post*, "Langley Field Airman to Get Cheney Award," January 21, 1928, 2.

418. "USS Akron Crash—Officers and Crew," Airships: The Hindenburg and Other Zeppelins, accessed August 22, 2015, http://www.airships.net/us-navy-rigid-airships/uss-akron-crash-officers-crew.

419. "John Mitchell *ROMA* Research Collection."

420. *Cumberland (MD) Evening Times*, "Hero of 1922 *ROMA* Air Disaster Dies," April 22, 1940, 3.

421. Mitchell, "The *ROMA* and her Crew," 7.

422. Mike Porter, "Virden T. Peek (1901–1982)—Find A Grave Memorial," Find A Grave—Millions of Cemetery Records, last modified August 9, 2010, http://www.findagrave.com/cgi-bin/fg.cgi?page=gr&GSln=peek&GSfn=virden&GSbyrel=all&GSdyrel=all&GSob=n&GRid=56835742&df=all&.

423. Johnson, Jouett and McNarney, "Proceedings," 57.

424. Colonel John D. Reardon (USAF, ret.), *COL John D. Reardon (USAF, ret.) in a Letter to Commandant, Air Force University, Maxwell Field*, (Asheville, NC: Letter Courtesy of Air Force Historical Research Agency, Maxwell Air Force Base, 1960).

425. Ibid.

426. Marvin & Samme Templin, "Augustine Biedenbach (1906–1961)—Find a Grave Memorial," Find a Grave—Millions of Cemetery Records, last modified November 30, 2011, http://www.findagrave.com/cgi-bin/fg.cgi?page=gr&GSln=biedenbach&GSfn=augustine&GSbyrel=all&GSdyrel=all&GSob=n&GRid=81263962&df=all&.

427. Marvin & Samme Templin, "Joseph M. Biedenbach (1900–1957)—Find a Grave Memorial," Find a Grave—Millions of Cemetery Records, last modified November 30, 2011, http://www.findagrave.com/cgi-bin/fg.cgi?page=gr&GSln=biedenbach&GSfn=joseph&GSmn=m&GSbyrel=all&GSdyrel=all&GSob=n&GRid=81263964&df=all&.

428. *Mt. Vernon (IL) Register-News*, "Airman from Illinois Killed," June 26, 1957, 1.

429. Judie Loudon, e-mail correspondence with the author, 2015.

430. St. Clair County, Illinois Vital Record: Medical Certificate of Death for Alberto Flores, St. Clair County, Illinois, 1988.

431. Certificate of Death, Commonwealth of Virginia, for Stella Saunders, Hampton, VA, Commonwealth of Virginia Department of Health, Bureau of Vital Statistics, 1949).

432. Mitchell, "The *ROMA* and Her Crew," 28.

433. Paul Hays, "Col. Byron T. Burt (1889–1969)—Find a Grave Memorial," Find a Grave—Millions of Cemetery Records, last modified December 6, 2013, http://www.findagrave.com/cgi-bin/fg.cgi?page=gr&GSln=burt&GSfn=byron&GSmn=t&GSbyrel=all&GSdyrel=all&GSob=n&GRid=121280858&df=all&.

434. *Washington Post*, "Gen. Walter J. Reed, 69; Pilot of Ill-Fated *ROMA*," July 12, 1963, B5.

435. *Hampton (VA) Daily Press*, "Area Deaths and Funerals: Mrs. Maria Reed Died in Richmond Hospital," March 12, 1976, 8.

436. Reed Jr., phone interview with the author.

437. Ibid.

BIBLIOGRAPHY

Aerocave *ROMA*—Italian Flight Log. Rome, Italy: Italian government, 1920.

Airship Heritage Trust. "Airships: R38." http://www.airshipsonline.com/airships/r38/ (accessed August 4, 2015).

Brooklyn (NY) Daily Eagle. "Airship *ROMA* Due to Reach Here Today." June 2, 1921, 3.

———. "Mitchel Field, as New Aerial Hub, to Get Dirigible *ROMA*." December 19, 1921, 2.

Certificate of Death, Commonwealth of Virginia, for Stella Saunders. Hampton, VA: Commonwealth of Virginia Department of Health, Bureau of Vital Statistics, 1949.

Chambers, Girard. Interview by Jennifer Doerr, 2005.

Charles W. Dworack: Report to Accompany H.R. 666. Washington, D.C.: Seventy-third Congress, Second Session, 1934.

Chicago Daily Tribune. "No More *ROMA* Type of Ships, Weeks Implies." February 24, 1922, 10.

Cincinnati Enquirer. "*ROMA* to Visit Akron." October 29, 1921, 4.

———. "Statement on *ROMA* is Reiterated." February 26, 1922, 26.

Contract No. 7105-B, Air Service Order No. 900023-B. Air Service, United States Army, 1921.

"The C-7 Tests." United States Naval Institute Proceedings 48 (January 1922): 305.

Cumberland (MD) Evening Times. "Hero of 1922 *ROMA* Air Disaster Dies." April 22, 1940, 3.

Curtis, Robert I., Martin Copp and John B. Mitchell. *Langley Field: The Early Years*. Newport News, VA: Langley Air Force Base, Office of History, 4500th Air Base Wing, 1977.

"Dirigible *ROMA* Volo di Consegna All Equipaggio Americano, Roma-Napoli, 15 Marzo 1921." Stabilimento di Costruzoni, Aeronautiche Laboratorio Fotografico. 1921. Courtesy of Fabio Iaconianni.

East Liverpool (OH) Evening Reviewer. "Open Probe of Crash of Dirigible *ROMA*." February 22, 1922, 1.

El Paso (TX) Herald. "Germans Send Sympathys." February 23, 1922, 7.

"Enduring Legacy: The Jamestown 1907 Exposition and the U.S. Navy." NHHC, http://www.history.navy.mil/museums/hrnm/interactive/1907exposition/index.htm (accessed August 3, 2015).

"Engines on the Airship *ROMA* to Be Changed." *Aerial Age Weekly*, January 1921, 499.

Fifty-third Annual Report of the Association of Graduates of the United States Military Academy at West Point, New York. West Point, New York: United States Military Academy at West Point, 1922.

Forrester, Leo C., Jr. "Remembering the *ROMA*." Researcher News Online. http://researchernews.larc.nasa.gov/archives/2002/022202/Roma.html (accessed August 10, 2014).

Fort Wayne (IN) Sentinel. "Cause is Mystery." February 22, 1922, 14.

Golson, Jordan. "WWI Zeppelins: Not Too Deadly, but Scary as Hell." Wired Online. Last modified October 3, 2014. http://www.wired.com/2014/10/world-war-i-zeppelins/.

Grossman, Dan. "Hindenburg Flight Operations and Procedures." Airships: The Hindenburg and Other Zeppelins. Last modified August 24, 2009. http://www.airships.net/blog/hindenburg-flight-ops.

Hall, Larry. "Crash of Dirigible *ROMA* Killed 34 in Norfolk in 1922: News." *Richmond (VA) Times-Dispatch*. Last modified February 18, 2009. http://www.richmond.com/news/article_7d4e8e1a-7eb1-5856-b610-bd44d021292a.html.

Hampton (VA) Daily Press. "Area Deaths and Funerals: Mrs. Maria Reed Died in Richmond Hospital." March 12, 1976, 8.

Handwritten notes on *ROMA*. Langley Air Force Base, VA: Air Combat Command, Joint Base Langley-Eustis, n.d.

Harrisburg (PA) Evening News. "King Emmanuel Sees Vatican From Air." October 28, 1920, 2.

Harrisburg (PA) Telegraph. "C.E. Diehl Has Piece of Fabric from *ROMA*." March 15, 1922, 13.

―――. "Knew *ROMA* Was Unsafe for Flight." February 23, 1922, 1.

Hays, Paul. "Col Byron T Burt (1889–1969): Find a Grave Memorial." Find a Grave—Millions of Cemetery Records. http://www.findagrave.com/cgi-bin/fg.cgi?page=gr&GSln=burt&GSfn=byron&GSmn=t&GSbyrel=all&GSdyrel=all&GSob=n&GRid=121280858&df=all&.

Hewitt, H.K. "Biggest Navy Blimp Burns with Three More." Proceedings—United States Naval Institute 47 (September–October 1921): 1,650.

History.com. "Battle of the Ironclads—Mar 09, 1862." http://www.history.com/this-day-in-history/battle-of-the-ironclads (accessed August 2, 2015).

"History." Fort Monroe Authority. http://www.fmauthority.com/about/fort-monroe/history (accessed August 3, 2015).

Hitchings, Robert. Personal interview with the author (Norfolk, VA), May 2014.

Hoffman, Earl. Letter to Stella Hoover, 1922. Battle Creek, MI: Courtesy of Air Combat Command History Office, Joint Base Langley-Eustis.

Hoffman, Nellie. Letter to Stella Hoover, 1922. Eaton Rapids, MI: Courtesy of Air Combat Command History Office, Joint Base Langley-Eustis.

Hoover, Stella. Letter to her mother, 1922. Eaton Rapids, MI: Courtesy of Air Combat Command History Office, Joint Base Langley-Eustis.

Hutchinson, William K. *Oakland (CA) Tribune*, February 22, 1922, 1.

Indianapolis (IN) News. "*ROMA*'s Battered Hulk Back at Langley Field." June 24, 1922, 2.

Indianapolis (IN) Star. "Member of Graveyard Club, Made Up of *ROMA*'s Crew, Predicted Death to Girl." February 24, 1922, 1.

"Inquiry into Operations of the United States Air Services: Hearing Before the Select Committee of Inquiry into Operations of the United States Air Services." United States Congress, Washington, D.C.: Sixty-eighth Congress, 1925.

John Mitchell *ROMA* Research Accn. #X.66. Hampton, VA: Hampton History Museum, n.d. Held in the archives of Hampton History Museum.

Johnson, Davenport, John H. Jouett and Joseph McNarney. Proceedings of a Board of Officers in Regard to the Accident to the Airship *ROMA*, February 21, 1922. Langley Field, VA: United States Army Air Service, 1922. Available on microfilm courtesy of Air Force Historical Research Agency, Maxwell AFB

Johnston, Vic. "Do You See What I See?" 1st Fighter Wing Public Affairs. http://www.jble.af.mil/shared/media/document/AFD-091215-069.pdf (accessed August 12, 2013).

Kansas City (MO) Star. "Prepare *ROMA* for Shipment." April 30, 1921, 31.

Kepner, Lieutenant General William E. Letter to historian John B. Mitchell. John Mitchell *ROMA* Research Accn. #X.66, Hampton History Museum, Hampton, VA, 1972.

Kolbert, James T. *Sheboygan (WI) Press Telegram*, February 22, 1922, 1.

"Langley History." NASA. https://www.nasa.gov/centers/langley/about/history.html#.Vci1svlVikp (accessed August 4, 2015).

Lawrence (KS) Daily Journal-World. "Regular Airflights." November 28, 1921, 1.

"Lieutenant General William E. Kepner." United States Air Force. http://www.af.mil/AboutUs/Biographies/Display/tabid/225/Article/106566/lieutenant-general-william-e-kepner.aspx (accessed January 5, 2015).

Logansport (IN) Pharos-Tribune. "America Buries *ROMA* Victims." February 23, 1922, 1.

Log of the ROMA. Langley Field, VA: U.S. Army Air Service, 1922.

Los Angeles Times. "*ROMA* Now Queen of Air." December 22, 1921, 12.

Loudon, Judie. Personal e-mail correspondence with the author, 2015.

Mabry, Dale. Dale Mabry to Commandant, Airship School, February 1, 1922. Langley Field, VA: From Air Force Historical Research Agency, Maxwell Air Force Base, 1922.

Mitchell, John B. *The Army Airship* ROMA. Hampton, VA: Syms-Eaton Museum, 1973.

———. "The *ROMA* and her Crew." *American Aviation Historical Society Journal* 18, no. 1 (Spring 1973): 1–7.

Mt. Vernon (IL) Register-News. "Airman from Illinois Killed." June 26, 1957, 1.

Muskogee County (OK) Democrat. "Many Thousand Warplanes Are Store [*sic*]—Useless." August 18, 1921, 1.

Muskogee (OK) Times-Democrat. "Inquiry to Fix Cause Is Begun in '*ROMA*' Crash." February 22, 1922, 10.

"Naval and Military Aeronautics." *Aerial Age Weekly* 13 (June–July 1921): 447.

"The New Italian Usuelli Semi-Rigid Airship." *Flight: The Aircraft Engineer & Airships*, September 1920, 44.

New York Times. "Envoy Johnson Takes 500-Mile Air Trip." March 16, 1921, 9.

———. "Giant Army Dirigible Wrecked." February 22, 1922, 2.

———. "Injured Survivors Tell of Disaster." February 22, 1922, 1.

———. "Lieut. Riley Elated at *ROMA* Assignment." February 22, 1922, 1.

———. "Fight Started for Funds to Develop Helium." February 23, 1922, 2.

———. "Sad Fabric of *ROMA* Was in 'Bad Shape.'" February 24, 1922, 1.

Nobile, Umberto. "Zeppelin Design: Umberto Nobile and the Italian N-1 Airship." Interview by Kenneth Leish, 1960.

Norfolk (VA) Ledger-Dispatch. "Broken Controls Caused *ROMA* Crash." February 22, 1922, 1.

———. "Legislatures See Waterfront from Deck of Special Steamer; View Scene of *ROMA* Disaster This Afternoon." February 22, 1922, 2.

———. "Survivor Tells of Airship's Crash." February 22, 1922, 1.

Norfolk (VA) Post. "'Better, Thank You, Sir!' *ROMA* Injured Say." February 24, 1922, 1.

Norfolk (VA) Times-Dispatch. "Cleaning Up Debris of *ROMA* at the Base." February 27, 1922, 1.

———. "Pay Last Tribute to *ROMA* Victims." February 25, 1922, 1.

Norfolk (VA) Virginian-Pilot. "Airman's Rose Falls on Bier of *ROMA* Head." February 25, 1922, 1.

———. "May Urge Helium Production For Use in Dirigibles." February 25, 1922, 1.

Oakland (CA) Tribune. "Congress Absolved by Aero Cheiftan." February 23, 1922, 1.

———. "Dirigible *ROMA* Destroyed by Explosion; Many Dead." February 21, 1922, 3.

———. "*ROMA* O.K. at Start." February 23, 1922, 1.

———. "*ROMA* Sailing Too Low, Italian Expert Claims." February 23, 1922, 1.

"Obituaries." *Fifty-Third Annual Report of the Association of Graduates of the Untied States Military Academy at West Point.* West Point, NY: United States Military Academy at West Point, June 1922, 89.

Office of History, Air Combat Command, Langley Air Force Base. *Langley: 1916–2003.* Langley Air Force Base, VA: Air Combat Command, Langley Air Force Base, 2003.

Ogden (UT) Standard Examiner. "Airship Makes Speedy Voyage." March 16, 1921, 2.

Parsons (KS) Daily Sun. "Witness Says *ROMA* Death Result Burns." February 25, 1922, 1.

Plester, Jeremy. "Weatherwatch: Fog Led to Destruction of 1921 Airship." Guardian. Last modified August 18, 2011. http://www.theguardian.com/news/2011/aug/18/weatherwatch-airship-fog-zeppelin.

Porter, Mike. "Virden T. Peek (1901–1982): Find a Grave Memorial." Find a Grave: Millions of Cemetery Records. Last modified August 9, 2010. http://www.findagrave.com/cgi-bin/fg.cgi?page=gr&GSln=peek&GSfn=virden&GSbyrel=all&GSdyrel=all&GSob=n&GRid=56835742&df=all&.

Reardon (USAF, ret.), Colonel John D. Letter to Commandant, Air Force University, Maxwell Field. Asheville, NC: Letter courtesy of Air Force Historical Research Agency, Maxwell Air Force Base, 1960.

———. The Semi-Rigid Airship *ROMA.* Letter courtesy of the Air Force Historical Research Agency, Maxwell Air Force Base, 1960.

Reed, (USAF, ret.), Lieutenant Colonel Walter J., Jr. Personal letter to the author, August 2014.

———. Phone interview, August 2014.

Reed, Lieutenant Walter J. Aircraft Log Book, Ansaldo. Langley Field, VA: U.S. Army Air Service, 1921.

————. Aviation Engine Log Book (*ROMA*). Langley Field: U.S. Army Air Service, 1921.

"Report on Accident to the Airship *ROMA*: Official Report Fails to Determine with Absolute Accuracy the Causes of the Accident." *Aviation*, August 1922: 148–52.

Richmond (VA) Times-Dispatch. "Colonel Saunders to Represent Government at Public Service." February 24, 1922, 1.

Roberts, Kenneth L. "Italy From a Dirigible Window." *Saturday Evening Post*, August 13, 1921.

"*ROMA* Is Ready." *Aerial Age Weekly* 14, no. 1 (1921): 244.

"*ROMA* Tragedy." National Museum of the U.S. Air Force. Last modified April 9, 2015. http://www.nationalmuseum.af.mil/Visit/MuseumExhibits/FactSheets/Display/tabid/509/Article/196943/roma-tragedy.aspx.

"Rome to Rio de Janeiro." *International Shipping Digest* 1, no. 1 (1919): 51.

Rouse-Bottom, Dorothy. Interview by Jennifer Doerr, 2005.

Rouse, Parke. "Doomed Airship Called Langley Home." *Daily Press.* Last modified May 28, 1989. http://articles.dailypress.com/1989-05-28/news/8905260143_1_roma-hampton-roads-crash.

San Francisco Chronicle. "Survivor of *ROMA* Has Lucky Star." February 22, 1922, 1.

Schenectady (NY) Gazette. "34 Men Roasted Alive in Airship *ROMA* Crashing on High Power Wire; Only 11 Out of 45 Passengers." February 22, 1922, 1.

Sheboygan (WI) Press Telegram. "Stop Building Dirigibles for U.S. Use." February 23, 1922, 9.

————. "Tells of Disaster." February 23, 1922, 9.

————. "Work Must Not Stop." February 23, 1922, 9.

Sixty-seventh Congress Congressional Record. Washington D.C.: United States Congress, 1922.

Spooner, Stanley. "The *ROMA* Flies Again." *Aircraft Engineer & Airships*, 1921.

St. Clair County, Illinois Vital Record: Medical Certificate of Death for Alberto Flores. St. Clair County, Illinois, 1988. Certificate provided to author courtesy of Mrs. Judie Loudon

St. John Erickson, Mark. "The World's Largest Lighter-than-air Ship Flew into History at Langley Field." Dailypress.com. Last modified November 15, 2013. http://www.dailypress.com/features/history/our-story/dp-the-first-flight-of-the-roma-from-langley-field-20131115-post.html.

"Stock Footage—Crew Members and Navigation Captain Dale Mabry During Flight of Airship ZDUS-1 Flies over Sky of Norfolk, Virginia." Historic Stock Footage Archival and Vintage Video Clips and Photo Images from CriticalPast. n.d. http://www.criticalpast.com/video/65675036617_airship-ZDUS-1_crew-members_ships-at-harbor_navigation-captain.

Templin, Marvin, and Samme Templin. "Augustine Biedenbach (1906–1961): Find a Grave Memorial." Find a Grave—Millions of Cemetery Records. Last modified November 30, 2011. http://www.findagrave.com/cgi-bin/fg.cgi?page=gr&GSln=biedenbach&GSfn=augustine&GSbyrel=all&GSdyrel=all&GSob=n&GRid=81263962&df=all&.

———. "Joseph M. Biedenbach (1900–1957): Find A Grave Memorial." Find a Grave—Millions of Cemetery Records. Last modified November 30, 2011. http://www.findagrave.com/cgi-bin/fg.cgi?page=gr&GSln=biedenbach&GSfn=joseph&GSmn=m&GSbyrel=all&GSdyrel=all&GSob=n&GRid=81263964&df=all&.

United States Holocaust Memorial Museum. "World War I: Treaties and Reparations." http://www.ushmm.org/wlc/en/article.php?ModuleId=10007428 (accessed August 4, 2015).

U.S. Army Air Service. "Images of the Reconstruction of *ROMA*." Courtesy of Air Force Historical Research Agency, Maxwell Air Force Base. 1922. Available through microfilm.

"The U.S. Army Dirigible *ROMA*: America's First Major Airship Disaster." *Gasbag Journal*, December/January 1999, 15.

U.S. National Archives. "Lighter Than Air, 1937." YouTube. n.d. https://www.youtube.com/watch?v=7hBZolnvWbQ.

"USS *Akron* Crash—Officers and Crew." Airships: The *Hindenburg* and Other Zeppelins. http://www.airships.net/us-navy-rigid-airships/uss-akron-crash-officers-crew (accessed August 22, 2015).

Van Nostrand, Major Percy E. "Lessons Learned from the '*ROMA*' Accident." *U.S. Air Services* 7, no. 1 (1922): 16, 28.

Washington Herald. "The Army's New Airship, *ROMA*, Nearly Rivals Lost ZR-2." August 30, 1921, 4.

Washington Post. "The Army's Airship." October 30, 1921, 35.

———. "Board Again Quizzes Survivors of *ROMA*." March 12, 1922, 4.

———. "Dance Celebrated Launching of Fatal Flight of the *ROMA*." February 22, 1922, 11.

———. "Gen. Walter J. Reed, 69; Pilot of Ill-Fated *ROMA*." July 12, 1963, B5.

———. "Langley Field Airman to Get Cheney Award." January 21, 1928, 2.

———. "*ROMA* Battles Gale." December 22, 1921, 1.

———. "*ROMA* Survivor and His Mascot." February 26, 1922, 6.

Watts, Maureen P., and Jakon Hayes. "The *ROMA* Airship Disaster Over Norfolk | HamptonRoads.com | PilotOnline.com." http://hamptonroads.com/2010/06/roma-airship-disaster-over-norfolk.

Winters, S.R. "America's Latest Airship—'*ROMA*.'" *Scientific American*, January 1922.

INDEX

T

U

V

W

ABOUT THE AUTHOR

Nancy E. Sheppard is a writer and historian of her native Hampton Roads, Virginia. She received her education in history from American Military University and Old Dominion University, specializing in the history of Hampton Roads from 1890 to present. This includes the influence of the military in the region. After publishing several short local history pieces online, she has devoted her research and writing over the past four years to telling the story of the U.S. Army Air Service dirigible *ROMA* and her crew.

Aside from history and writing, she is a tireless advocate for the awareness of Autism Spectrum Disorders as well as for special needs family members in the military community. In her free time, she enjoys reading, writing, art and spending time with her husband and their two children.

Please visit her website (http://www.nancyesheppard.com) or follow her on Facebook (https://www.facebook.com/nancyesheppardauthor).